CERN AND THE HIGGS BOSON

OTHER HOT SCIENCE TITLES AVAILABLE FROM ICON BOOKS

Destination Mars:
The Story of Our Quest to Conquer the Red Planet

Big Data:
How the Information Revolution is Transforming Our Lives

Gravitational Waves:
How Einstein's Spacetime Ripples Reveal the Secrets of the Universe

The Graphene Revolution:
The Weird Science of the Ultrathin

CERN AND THE HIGGS BOSON

The Global Quest for the Building Blocks of Reality

JAMES GILLIES

Published in the UK and USA in 2018
by Icon Books Ltd, Omnibus Business Centre,
39–41 North Road, London N7 9DP
email: info@iconbooks.com
www.iconbooks.com

Sold in the UK, Europe and Asia
by Faber & Faber Ltd, Bloomsbury House,
74–77 Great Russell Street,
London WC1B 3DA or their agents

Distributed in the UK, Europe and Asia
by Grantham Book Services,
Trent Road, Grantham NG31 7XQ

Distributed in the USA
by Publishers Group West,
1700 Fourth Street, Berkeley, CA 94710

Distributed in Australia and New Zealand
by Allen & Unwin Pty Ltd,
PO Box 8500, 83 Alexander Street,
Crows Nest, NSW 2065

Distributed in South Africa
by Jonathan Ball, Office B4, The District,
41 Sir Lowry Road, Woodstock 7925

Distributed in India by Penguin Books India,
7th Floor, Infinity Tower – C, DLF Cyber City,
Gurgaon 122002, Haryana

Distributed in Canada by Publishers Group Canada,
76 Stafford Street, Unit 300
Toronto, Ontario M6J 2S1

ISBN: 978-178578-392-0

Typeset in Iowan by Marie Doherty

Printed and bound in Great Britain
by Clays Ltd, Elcograf S.p.A.

CONTENTS

About the author vi
Author's note vii

1 Breaking news 1
2 Atomos 13
3 From the ashes 30
4 A new laboratory is born 44
5 The birth of the big machines 52
6 In theory 64
7 New kids on the block 81
8 From a November revolution to W and Z 99
9 The race for the Higgs 110
10 Supercollider! 126
11 What's the use? 144

Epilogue: What next? 153
Further reading 157
Index 159

ABOUT THE AUTHOR

James Gillies began his career at CERN as a graduate student in 1986. After eight years in research and a brief stint at the British Council in Paris, he joined the laboratory's communications group in 1995, heading the group from 2003 to 2015. He is now a member of CERN's strategic planning and evaluation unit.

AUTHOR'S NOTE

This book gives just a glimpse of the fantastic journey of discovery that is particle physics. It would be impossible in a book of this kind to tell the whole story, with all its twists and turns, dead ends and new beginnings. As a result, there are whole areas of physics, giants of the field, technological advances and major laboratories that are missing or only hinted at. Instead, I have focused on the electroweak physics to which CERN has contributed so much over the years, and included just the physics necessary to tell the story of the Higgs. I have tried to give an idea of how extraordinary it is that human intellect has delivered the theories and the machines that allow us to understand the workings of the universe at such an intricate and intimate level. There are those who say that science of this kind diminishes nature's beauty. On this point, I can only concur with the great Richard Feynman who offered the opposite view: science can only add to our sense of wonder. I hope that I have managed to convey some of that wonder in these pages.

I would like to thank Austin Ball, Stan Bentvelsen, Tiziano Camporesi, Dave Charlton, Jonathan Drakeford, Rolf Heuer, John Krige, Mike Lamont, Michelangelo Mangano, John Osborne and David Townsend, all of whom know much of this story far better than I, and generously gave up their time to read and improve the draft. Any remaining errors are my own. My thanks also go to series editor Brian Clegg, along with Duncan Heath and Robert Sharman at Icon Books for their many constructive comments and sensitive editing of the manuscript. Finally, I would like to thank my wonderful family for their patience and support.

BREAKING NEWS 1

15 June 2012

It was mid-afternoon when the phone rang. I was in the garden searching among weeds for the vegetables I'd planted a couple of months earlier. 'What I've just seen is not going away,' said the voice on the other end. It was Austin Ball, an old friend from the days when we were both working on the OPAL experiment at CERN, the European particle physics laboratory near Geneva. Earlier that day he'd seen the results of his experiment's search for the elusive Higgs particle. He'd been in the room when the physicists working on one of the big LHC (Large Hadron Collider) experiments had taken their first look at their results, and what they had seen had set hearts racing.

Such moments are few and far between: occasions on which a scientist, or in this case a roomful of scientists, can be the first to know something completely new to humankind. What I would have given to be in that room – but I'd traded my research career 20 years ago for a job in CERN's public communications team. Austin had thought long and hard before calling me, and for good reason: new results are closely guarded secrets until the experimenters are absolutely sure they are ready to go public. I felt honoured to be trusted enough to be brought into this privileged inner circle; and now, sworn to secrecy, I knew we had a job to do. We had to get ready for the biggest announcement in the laboratory's history. And we had to do it with the utmost discretion.

The Large Hadron Collider is CERN's flagship research instrument. It had risen to notoriety some four years earlier for all kinds of reasons. As the world's largest scientific instrument, with a price tag to match, and host to global collaborations involving thousands of scientists and engineers of around 100 nationalities, it had grabbed the popular imagination. For many, CERN's quest to understand the weird and wonderful universe we inhabit represented the true spirit of humanity; a model of what people can do when they put aside their differences and work together to achieve a common goal. To others, however, it was irresponsible, dangerous, or even redolent of the biblical story of Babel: an arrogant affront to the divine.

Whatever people thought, the net result was that the eyes of the world were on CERN, and when the time came to announce this particular result, it would not be a quiet affair in front of an exclusive audience of physicists in the lab's main auditorium. This would be much bigger.

Timescales are long in particle physics. The LHC was first imagined in the late 1970s, and one of its main research goals went back even further, to 1964. That was the year that Robert Brout and François Englert, and independently Peter Higgs, published papers in the journal *Physical Review Letters* proposing a mechanism that would give mass to fundamental particles. Why should anyone care about that? Because we, and everything we can see in the universe, are made of fundamental particles, and without mass those particles would be unable to form anything solid. In other words, we would not exist.

From the early 1960s, understanding mass ranked among the most pressing of riddles in fundamental physics, and it would take almost half a century to solve. Thankfully, physicists are usually blessed with a great deal of patience. Before any experiments would be ready to deliver the experimental evidence to confirm the idea of Brout, Englert and Higgs, a decade of theoretical work would be needed. It would be several decades before technology delivered the instruments that would eventually crack the enigma.

Research and development for the LHC began in the mid-1980s, while experimental collaborations started to form in the early 90s. The project was fully approved by 1996, and construction began soon after. By 2008, the machine was ready to go, and under the eyes of the global media, a beam of particles was circulated for the first time on 10 September 2008. It was a day of great excitement at CERN. 'Just another day at the office, eh?' said LHC project leader Lyn Evans as I headed for home at the end of the day. But the elation was short lived. Just nine days later, the LHC suffered a setback from which it would take a year to recover: a helium leak led to extensive damage to the machine. Meanwhile, at Fermilab in the United States, another remarkable particle collider, the Tevatron, a venerable machine first switched on in 1985, was limbering up for one last push to discover the particle that had come to be known as the Higgs. Discovering the Higgs particle would bring confirmation that Brout, Englert and Higgs were right. The race was on.

Although particle theory was very clear that a mechanism for mass was needed, and would have to appear at the particle collision energies of the LHC, there was one key feature of the Higgs that it did not predict: the particle's mass. It could well be in range of the Tevatron – nobody knew. But if the Higgs existed at all, it would definitely be in range of the LHC. The currencies of particle physics are mass and energy, with the exchange rate being the speed of light squared. That's what Einstein's famous equation $E=mc^2$ tells us, and it's why particle accelerators concentrate energy in a tiny space, converting it to mass in the form of new particles. The higher the energy of the accelerator, the higher the mass of the particles that can be produced, and the LHC was designed for a collision energy some seven times higher than that of the Tevatron.

By the end of 2009, the LHC was back in the race – and with a vengeance. Records rapidly fell, and on 30 March 2010, data collection began. The days of sudden realisations leading to 'Eureka!' moments in fundamental physics research are long gone. In modern particle physics research, discovery often comes through a painstaking analysis of vast quantities of data, looking for subtle signals that known physics can't explain. Like everything else in modern research, Eureka requires patience.

Data came rolling in fast as the LHC performed better and better, but nobody was looking at what the data were saying. The main analyses run shielded from the view of human eyes until the time is deemed right to take a look. The reason that scientists do this is that humans are very good at seeing things that aren't really there, and then skewing their interpretations to match their preconceptions. Algorithms know no such bias, and can be trusted to conduct the analyses free from prejudice.

Nervous eyes were scanning the horizon for hints of what might be happening across the Atlantic at Fermilab, but all was quiet there as well. By spring 2011, combined analyses from CERN and Fermilab had shown where the Higgs particle was not. They had narrowed down the range of masses it could have to 114–157 GeV, with a small window up at 185 GeV. A GeV – or Giga electron Volt – is a unit of mass used in particle physics. In everyday terms, it's tiny. There are over 500 billion trillion GeVs in 1 gram. But in the world of fundamental particles, the Higgs, if it existed at all, would be a very heavy thing. To put it in context, the basic building blocks of atomic nuclei, protons and neutrons, weigh in at just about 1 GeV, and by the time we reach 185 GeV, we're looking at atoms of heavy metals like tungsten.

In 2011, the Higgs was running out of places to hide, and everyone in the global particle physics community knew that representatives of the two LHC experiments spearheading the search, named ATLAS and CMS, would have to say something at the big summer conference in Mumbai, and so would their rivals at D0 (dee-zero) and CDF at Fermilab.

The conference came and went with no discovery in sight, and analysis continued apace. On 13 December, CERN organised a Higgs Update seminar to satisfy the demand for information coming from the global physics community. About ten times more people tuned into the webcast than there are particle physicists in the world, and they learned that the mass range for the Higgs had been squeezed to just 115–130 GeV, with both CERN experiments reporting that they might be seeing hints of something new hiding among the data with a mass of about 125 GeV. The signal was too weak for the experiments to be sure, but there was a new sense of excitement in the air. It was tantalising, but everybody was trying not to get too excited. Hints of new physics come and go, but as the seminar drew to a close, someone made the comment that if the Higgs existed, we'd know next year.

On 5 April 2012, the LHC resumed running, this time as the world's only high-energy particle collider. Fermilab's Tevatron had collected its last data on 29 September 2011, and although the D0 and CDF analyses were still ongoing, it was beginning to look like it would be up to the LHC experiments to prove the existence of the Higgs. Eyes were focusing ever more closely on CERN.

Spring and early summer were the calm before the storm. Data were coming in and analyses were running, but news from the experiments was scarce. Not only do analysis teams run their analyses blind, they also keep them to themselves for as long as possible to ensure that each analysis is independent. They do this because reproducibility is vital for science: if one experiment sees something and another does not, the chances are that someone's made a mistake in their analysis, but if two completely independent experiments see the same thing, the chances are that it's real.

Each experiment was reporting progress individually to CERN's Director General, and as the summer conference season again approached, we had to decide what to do. That was when my old friend from OPAL, by this time working on the CMS experiment, disturbed me from my gardening.

22 June 2012

The International Conference of High Energy Physics (ICHEP) was scheduled to run from 4–11 July 2012 in

Melbourne, Australia, and the party line from CERN that spring was that if a discovery were to be announced, we'd do it at CERN; anything else would be reported at ICHEP. The way things looked, we were working towards the latter option. Despite the big time difference, plans were put in place to relay the presentations back from Melbourne to CERN's main auditorium so scientists there could take part. The media were clamouring for news of what was going to be said, but there was nothing to say. Even after what Austin had seen, if a discovery were to be announced, the experiments would need to be absolutely certain, and time was running out.

It looked as though the summer conference would come just a little too soon for 2011's tantalising hints to crystallise into a strong enough signal to announce a discovery, and the CERN communications team started looking forward to a relatively tranquil summer. But then our plans were thrown into disarray. When members of the CERN Council gathered for their regular summer meeting on 21–22 June, they declared that whatever was to be announced, it would be announced at CERN. We issued a press release to that effect on 22 June – 'CERN to give update on Higgs search as curtain raiser to ICHEP conference' – fuelling speculation that a discovery announcement was on the cards.

The Melbourne plan was rapidly turned on its head. Instead of relaying the conference sessions to CERN, a second Higgs Update seminar would be held at CERN on 4 July – the only day compatible with CERN's agenda and that of the conference – and it would be relayed to conference delegates arriving in Melbourne. Why had CERN changed its mind, people began to ask? Surely they'd only do that if they had a major announcement to make. The truth of the matter is that we still did not know whether there would be enough to announce a discovery or not.

Some people were taking no chances. On 26 June, I received an email from Carl Hagen asking whether he and his colleague Gerry Guralnik could attend. Hagen and Guralnik, along with the British physicist Tom Kibble, had conducted early work on the mechanism of mass in the 1960s, independently of Brout and Englert and of Higgs. We replied that they'd be welcome. The following day, I wrote to François Englert, Peter Higgs and Tom Kibble inviting them to attend. Robert Brout had sadly passed away on 3 May 2011. Higgs and Englert said they'd be delighted, while Kibble replied that he'd be attending a Higgs update event organised in Westminster to which the British Prime Minister and Science Minister had been invited.

The big day was fast approaching and still Rolf Heuer, the Director General of CERN, did not know whether he'd be presiding over a major announcement or another cliff-hanger. By this time, each experiment was preparing to freeze its analysis for ICHEP, and the Director General had seen them both. Suddenly, it came to him: even if neither experiment was able to claim the five sigma significance required to announce a discovery, he knew that they were both close enough that when the data were combined, it would pass the threshold. He made the decision – it was going to be a discovery announcement.

Sigma: one small word that means a great deal to particle physicists in search of a discovery. Sigma gives a measure of the statistical significance of a measurement. In other words, it's the likelihood that what appears to be a real phenomenon could just be the result of pure chance. For example, one sigma corresponds to a 32 per cent probability of a statistical fluke, two sigma 5 per cent, and three sigma 0.3 per cent.

Now imagine rolling a die and getting sixes ten times in a row. It's unlikely, but possible. By the time you get 100 sixes in a row, it's looking increasingly likely that your die is loaded, but it could still be down to pure chance. It's the same in particle physics, with a five sigma measurement being one that's so unlikely to be down to chance that physicists feel comfortable to cry Eureka! In this case, five sigma corresponds to a chance of just one in about 3.5 million that the observed events are a statistical fluctuation and not a signal for the existence of the Higgs.

Things were moving fast. The science press started to call, trying to get a scoop. Some had been offered results from anonymous sources and were seeking confirmation from the CERN press office. They would have to wait. There was a physics summer school in the Sicilian hilltop town of Erice. Stan Bentvelsen, the Director of the Netherlands' national laboratory for particle physics was there and so was Peter Higgs. Stan was a member of the ATLAS collaboration, and he was there to make a documentary film about particle physics with a Dutch filmmaker. In one scene, he's pictured showing Higgs the already-published results from ATLAS. Soon after, Higgs got an invitation to a seminar at CERN. On the basis of what Stan knew, but was not saying, he told Higgs: 'Take this invitation seriously and go.'

On 2 July, the CDF and D0 experiments published the final word on Fermilab's search for the Higgs. They reported tantalising hints in the same mass range as the LHC experiments, but only at around the three sigma level, not nearly enough to be sure. All eyes were now firmly on CERN. On 3 July, the eve of the second Higgs Update seminar, a video interview of Joe Incandela, the CMS experiment's spokesperson, was inadvertently made visible for just a few minutes on the CERN website. We'd recorded two interviews: one for a discovery, the other for a cliffhanger, and the plan was to use the appropriate one on the day. This was the discovery one, and the few minutes it was visible were all that were needed for half a dozen or so keen-eyed journalists to see it. Was it real, they asked? Wait until morning, we said.

The big day arrived. Those who had camped out overnight to ensure their place in the CERN auditorium rolled up their sleeping bags and waited for the doors to open. François Englert, Gerry Guralnik, Carl Hagen and Peter Higgs were ushered to their seats, while the crowds of journalists who had made the trip to CERN were shown to CERN's council chamber, in which a press conference would follow, to watch the seminar on a big screen.

There's a curious dichotomy in science. Papers begin with an abstract that summarises the main points of the paper, and then go on to explain in painstaking detail how the scientists got there. Scientific talks do it the other way round, taking the listener through step by step while saving the conclusion to the end. The auditorium was on tenterhooks, along with those gathered in Melbourne and the 500,000 people watching the webcast. By the time the ATLAS and CMS spokespersons, Fabiola Gianotti for ATLAS and Joe Incandela for CMS, concluded, each of the experiments reporting a five-sigma signal, Rolf Heuer declared: 'As a layman, I would now say I think we have it. You agree?' The deafening applause spoke for itself.

Somewhat ironically, by the time the CERN audience learned of the discovery, it was already headline news around the world. The UK Science Minister had turned up to the aforementioned event in Westminster, and John Womersley, the Chief Executive of the UK's Science and Technology Facilities Council, decided not to keep him waiting. News of the Higgs discovery was broken in London, not Geneva after all, but for those in the CERN auditorium that day, there was no better place in the world to be.

The image of the day was of Peter Higgs wiping a tear from his eye as he learned that the idea he'd published half a century before had finally been shown to be right. 'For me it's really an incredible thing that it's happened in my lifetime,' he said. The emotion was no less profound for Carl Hagen and Gerry Guralnik and was perhaps all the more so for François Englert since his lifelong intellectual partner, Robert Brout, had not lived to see the day. As we escorted Peter Higgs to the council chamber for the press conference, he was jostled like a rock star and CERN's press officers had to take on the role of bodyguards. When asked for comment, he was a picture of magnanimity, declaring that this was a day for the experiments. He was right: the theorists' time would come, as he and Englert, who met for the first time at CERN on 4 July 2012, were invited to Stockholm the following year to receive the Nobel Prize.

Among the acres of media coverage of the discovery was a very poetic piece by Jeffrey Kluger in *Time* magazine that started with the line: 'If physicists didn't sound so smart you'd swear they were making half this stuff up', and concluded with a wonderful summary of what had happened that day. 'The boson found deep in the tunnels at CERN goes to the very essence of everything,' Kluger wrote. 'And in a manner as primal as the particles themselves, we seemed to grasp that, despite our fleeting attention span, we stopped for a moment to contemplate something far, far bigger than

François Englert (left) with Peter Higgs at CERN on 4 July 2012.

CERN

ourselves. And when that happened, faith and physics – which don't often shake hands – shared an embrace.'

To get to the heart of what it was that caused the world to stop that day, we need to go back in time, all the way to the Greek city of Miletus in the 5th century BC, to Democritus, and perhaps his master Leucippus. Democritus is widely credited as being the originator of the concept of atomism: the notion that there's a smallest indivisible particle of a substance that is still identifiable as being that substance. And we need to revisit Isaac Newton in Cambridge in the 1680s, making the first inroads into understanding the basic forces of nature at work between those 'atoms' of matter. It's a journey that spans centuries and continents and is a testament to the power of human ingenuity.

ATOMOS 2

Particles and forces

Take a piece of stuff – a pencil lead made of pure carbon, for example – and cut it in half. Is what remains still carbon? Do it again, and again, and again, each time asking yourself the same question. Is there a point at which you arrive at the smallest, indivisible object that can still be called carbon? This is the kind of question that the ancient Greek philosopher Democritus was asking himself in 5th century BC Athens. In the process, he established the notion of atomism. You could argue that he also unwittingly thereby established the field of particle physics, because particle physics is all about exploring the tiniest constituents of matter from which the universe and everything in it is made, along with the forces at work between them.

Particle physics aims to understand what it is that everything is made of, including ourselves, and why matter behaves the way it does. Why does it organise itself into things like Democritus' atoms at tiny distance scales? Why does it form complex objects such as tables, chairs, human beings and planets at intermediate distance scales, and why does it form structures like solar systems and galaxies at larger distance scales? To address these questions, particle physics, which studies the very small, teams up with cosmology, the science of the very large. This book will concentrate on the tiny.

The word atom derives from the Greek *atomos*, which simply means things that can't be cut into smaller pieces. It's not only in ancient Greece that the concept arose. References appear in other philosophical traditions, each postulating that everything we see and experience is made up of different configurations of a relatively small number of basic building blocks existing in what is otherwise a void.

The basic idea of the early atomists has stood the test of time. What we now call an atom of carbon is indeed the smallest indivisible object that can still be called carbon, though today's atoms are far from indivisible. Each is made up of a set of smaller particles arranged in particular configurations to make up the range of atoms we know.

In 1879, the Russian scientist Dmitri Mendeleev placed all the known atoms in order of their mass in his now famous periodic table of the elements. In doing so, he found that they formed groups of atoms sharing the same properties within the table. Gaps in the table allowed him to forecast the existence of yet-to-be discovered atoms, and to predict their properties. Over time, the holes were duly filled as the missing elements were found.

The fact that patterns appeared in the table suggested that underlying the atoms was some deeper structure. Mendeleev's periodic table pointed towards the atoms being composed of constituent particles, and it would not be long before the first fundamental particle of matter was discovered. The modern field of particle physics was about to be born.

In 1906, the British physicist Joseph John Thomson was awarded the Nobel Prize in physics 'in recognition of the great merits of his theoretical and experimental investigations on the conduction of electricity by gases'. In other words, he had discovered a particle called the electron.

In announcing the prize, the President of the Royal Swedish Academy of Science referred to the supposed existence of an elementary charge that had been postulated ever since Michael Faraday had shown in 1834 that every atom carries a charge that's a multiple of that of the hydrogen atom. Here, the word atom is used in its modern day sense to mean the smallest indivisible piece of any chemical element, but it was not long before people were referring to an 'atom of electricity': a smallest indivisible unit of electricity.

In experiments carried out at the Cavendish Laboratory in Cambridge in 1897, Thomson had quantified Faraday's atom of electricity. In his experiments, he passed an electric current through a glass tube with the air pumped out. The end of the tube was coated with a fluorescent material, on which a glowing dot appeared. This phenomenon was already known, and attributed to mysterious cathode rays. By applying electric and magnetic fields to the tube at the same time, Thomson caused the glowing dot to move, and when the effects of the two fields cancelled each other out, leaving the dot's position unchanged, he applied the equations of motion of cathode rays in electric and magnetic fields to work out the charge to mass ratio of the particles making up the cathode rays. His conclusion was revolutionary. These particles were much smaller than an atom, and he postulated that they emanated from the atoms themselves. Thomson had shown that atoms were not the fundamental building blocks of matter that people supposed them to be, but were themselves composed of smaller things. Thomson's discovery not only opened up a new field of research, it also led to new devices ranging from oscilloscopes to televisions, which, until recently, used cathode rays to trace out patterns on the screen.

The discovery of other fundamental particles came thick and fast. Ernest Rutherford, working in Manchester, discovered alpha particles in the radioactive decay of uranium in 1899. Alpha particles would later turn out to be the nuclei of helium atoms, composed of two protons and two neutrons, and Rutherford was quick to put them to use. In a 1911 experiment carried out by his graduate students Hans Geiger and Ernest Marsden, alpha particles were used to bombard a gold foil to test the notion that atoms were essentially like plum puddings – balls of positive charge with the negatively charged electrons dotted around like raisins.

What Geiger and Marsden saw turned that notion on its head. They found that most of the time, the alpha particles just passed right through the gold foil as if it were not there, but occasionally they would bounce straight back. This suggested that rather than being a uniformly distributed blob of matter, atoms had most of their mass concentrated in the middle, with the electrons orbiting rather like planets around a star. Geiger and Marsden had discovered the atomic nucleus, and by analysing the precise way that the alpha particles scattered, were able to work out that the nucleus is very tiny. It is the orbits of the electrons that determine an atom's size, and atoms are about 99 per cent empty space.

Rutherford went on to discover the positively charged proton in experiments carried out between 1917 and 1919, and when James Chadwick discovered its electrically neutral counterpart, the neutron, in 1932, the underlying structure suggested by the periodic table of the elements was complete. All the great diversity of the atoms is accounted for by differing combinations of just three particles: protons, neutrons and electrons. The simplest atom, hydrogen, consists of a single proton orbited by a single electron, whereas something heavy like the gold in Rutherford's experimental foil has 79 protons and electrons, with 118 neutrons in the nucleus.

The complexity of the periodic table had been simplified to just three building blocks, but that was not the end of the story. Other particles were being discovered in the early part of the 20th century. When Paul Dirac found two possible solutions to his equation for the electron in 1928, one with a negative charge, the other positive, he boldly predicted the existence of a positively charged electron. The positron duly made its appearance in an experiment performed by Carl Anderson in California in 1932, becoming the first particle of antimatter to be observed. The young field of particle physics was thriving, and the seed for countless science fiction stories had been sown: where would the USS Enterprise and its antimatter drive be without Dirac?

Another particular anomaly was suggested in the 1920s through the radioactive decay of certain elements. At that time, the form of radioactivity known as beta decay was understood to be due to a nucleus shedding an electron by turning one of its neutrons into a proton, but there was a problem. Beta decay is very common – one example that anyone who has ever eaten a banana will have been exposed to is the beta decay of the potassium isotope, potassium-40, which is found in many foods. It decays to the stable isotope calcium-40. The problem is that beta emitters emit electrons not at a constant energy, as inviolable conservation laws would require if there were only two particles involved (the nucleus and the electron), but with a continuous spectrum.

In 1930 Wolfgang Pauli proposed a solution in an open letter sent to Lise Meitner for consideration at a conference she was involved with in Tübingen. Meitner was one of the great physicists of her time, and would go on to discover the fission of heavy elements with Otto Hahn, work for which he would later be awarded the 1944 Nobel Prize in chemistry. Although the omission of Meitner remains controversial, when the Nobel committee's archives for that year were opened up in the 1990s, it appeared that the omission was due to a lack of understanding by the chemistry prize committee of the role she had played in elucidating the underlying physics. Nevertheless, there has always been a broad consensus that whatever the reason for her omission, she should have had a share of the prize.

Pauli began his letter: 'Dear radioactive ladies and gentlemen', and went on to propose that the peculiar observations from beta decay could be accounted for if the energy were shared between two particles, an electron and an ethereal, yet to be discovered, neutral particle, which he called a neutron. Three years later, Enrico Fermi gave it the name neutrino, on the basis that neutron was already taken, but it would be many years before Pauli's hypothetical particle would be discovered.

Over the following decades, a range of particles was discovered in the cosmic radiation that is constantly bombarding the Earth from space. One, the muon, was so unexpected that it prompted the American physicist, Isidor

Rabi to quip: 'Who ordered that?' Rabi would go on to play a key role in the creation of the European Organization for Nuclear Research, CERN.

How all these newly discovered particles fit into the story of matter would remain perplexing for decades, but it was beginning to look as though another periodic table would be needed: a periodic table of the particles that could point the way to a deeper structure of matter. One thing, however, was certain. Particle physicists would not be getting bored any time soon.

Collecting new particles like stamps in an album is not the only thing needed to understand nature at the smallest distance scales. It is also necessary to understand what it is that causes the particles of matter to behave the way they do. What is it that causes protons and neutrons to clump together to form nuclei, and why do electrons join them to make atoms? What is it that causes atoms to stick together to form molecules and complex objects like crystals, or human beings? And what is it that causes matter to organise itself into stars and solar systems, galaxies and galactic clusters at the largest distance scales? In short, what is it that choreographs the cosmic ballet from the smallest distance scales to the largest?

The answer lies with the fundamental forces of nature, and to understand them, it's customary to take a trip back to a garden in Lincolnshire in 1666. There legend has it that Isaac Newton, who had been forced to leave Cambridge because of the plague, was contemplating the motion of the Moon around the Earth when a falling apple inspired him to one of the most remarkable strokes of genius in the history of science: the force that causes apples to fall to the ground is the same as that that causes the Moon to orbit the Earth, and indeed the Earth to orbit the Sun. It is a force that must operate over vast distances, and it decreases with the inverse square of distance. Over the next few years, Newton elaborated his theory of gravity and published it in the *Philosophiae Naturalis Principia Mathematica* in 1687.

This was a remarkable piece of work, and has served humanity well. Newton applied his theory of gravity to account not only for the motion of the Moon around the Earth, but also to explain the tides, to predict the relative masses of the giant planets in the solar system, and to account for the eccentric orbits of comets. In recent times, Newtonian mechanics has been used to calculate the trajectories of spacecraft, and to put people on the surface of the Moon.

Among the main aims of particle physics is the quest for simplicity. Wherever there appears to be complexity, physicists look for patterns that could point to an underlying simplicity. That's what happened with Mendeleev's periodic table, and was starting to happen again with the classification of the plethora of particles that were being discovered, starting with the electron in 1897. This is not only true for particles, but also for the forces that act between them. In realising that the force that causes apples to fall to the ground, and that keeps us firmly attached to planet Earth, is the same as the one responsible for celestial mechanics, Newton had taken the first step towards the unification of the forces into a single mathematical formulation. He'd started a journey that is still unfolding, and that is perhaps the most important outstanding question in physics today.

Nature is full of phenomena that appear to be independent, yet are in fact linked. Newton had made that jump with gravity, and almost two centuries later, the Scottish physicist James Clerk Maxwell would make a similar connection between electricity and magnetism. The signs were already there: magnets had long been shown to influence or even produce electric currents, but nobody had managed to write down a set of equations that explained both electricity and magnetism in a single theory. Maxwell did that in 1861, thereby taking a second step along the path to unifying the forces. It would be a further century before the next step was taken.

Relativity and quantum mechanics

In 1905, fundamental physics changed forever thanks to a series of papers published by a patent clerk in Bern. This was Albert Einstein's so-called miraculous year.

Einstein was married to Mileva Marić. They had trained together in Zurich in physics and mathematics, and her prowess probably contributed to Einstein's four papers that changed the course of physics. One paper concerned the seemingly random motion of particles suspended in a liquid, and it revolutionised the field of statistical mechanics. Another dealt with the interaction of photons, particles of light, with electrons. Photons above a certain energy could liberate electrons from materials, a phenomenon known as the photoelectric effect. This paper won Einstein the Nobel Prize in physics, and paved the way to the development of quantum mechanics, which governs the behaviour of matter at very small distance scales.

The third and fourth papers introduced the concept of special relativity, which showed the relationship between space and time, and established that the speed of light, denoted by c, is a cosmic speed limit determining the relationship between mass and energy through the best known equation in physics: $E=mc^2$. As we have seen, this remarkable equation tells us that energy, E, and mass, m, are interchangeable, with the exchange rate being the speed of light squared. That is an enormous number, which means that a small amount of matter can be exchanged for a large amount of energy, or that from a large amount of energy, it's possible to generate a small amount of mass. $E=mc^2$ would go on to become one of the foundation equations of experimental particle physics, being applied by researchers to produce new particles by concentrating energy through the collisions of other particles.

Einstein's paper on special relativity also paved the way for a much bolder theory, that of general relativity, which he published in 1915. This is Einstein's theory of gravity, which needed to reproduce the success of Newtonian mechanics within the four-dimensional geometry of space and time required by special relativity. In it, he describes not only the phenomenon of gravity, but also the reason for it: curved spacetime. Mass, said Einstein, deforms spacetime, thereby influencing the trajectory of anything that moves. In science centres the world over, this is demonstrated by placing heavy spheres representing planets on top of sheets of rubber, while rolling smaller objects around before and after to show how their trajectories change. Like many demonstrations, this one has its limits: it works for the two-dimensional space of the rubber sheet, but in reality, massive objects deform three-dimensional space, as well as time.

In 1919 a solar eclipse visible from South America to Africa provided an opportunity to test general relativity. If Einstein was right, light coming from stars in the constellation of Taurus would be bent by the Sun, making them visible in the shadow of the eclipse, while Newtonian gravity would have them hidden behind the Sun. British physicist

Arthur Eddington mounted expeditions to northern Brazil and the island of Principe off the coast of West Africa to observe the eclipse, and he found that the stars from Taurus appeared exactly where general relativity predicted they would. General relativity had its experimental confirmation as the theoretical framework describing the behaviour of the universe at very large scales of distance.

Several decades later, general relativity would go on to have an important practical consequence. While Newtonian mechanics works well enough to put people on the Moon, it's not sufficiently precise for the GPS systems that so many of us now take for granted. For that, Einstein's equations are needed to deliver the pinpoint accuracy we've come to take for granted.

While Einstein famously moved away from quantum mechanics, refusing to believe that God plays dice with the universe, his Nobel Prize-winning paper nevertheless inspired others. The Danish physicist Niels Bohr pioneered the development of quantum mechanics as the theory that underpins the interactions between particles. Quantum mechanics involves dividing things into very small quantities, rather than placing them on a continuum. It's a bit like the transition from analogue to digital, but at the level of fundamental particles and forces of nature. In quantum mechanics, there is a smallest indivisible piece of everything. For light, it's the photon, for electricity, a unit of electric charge. The centuries-old musings of Democritus were turning out to be more profound than he could ever have imagined.

For atoms, the consequence was that electrons could not simply orbit anywhere around the nucleus; the mathematical logic of quantum mechanics dictated that they had to be in a series of discrete orbitals, and never in between. Another profound consequence of quantum mechanics is that nothing is precisely knowable; the more precisely you measure one aspect of a system, the less well you can know another. This is Heisenberg's uncertainty principle, which quantifies the degree to which it is possible to measure correlated properties of a system. For example, if we measure the position of a particle precisely, we can't pinpoint its momentum. Conversely, if we measure its momentum, we cannot know precisely where it is at that time.

A range of counterintuitive thought experiments emerged, most famously that of Schrödinger's cat, sealed inside a box with a vial of poison, and existing in an uncertain state, neither alive nor dead, until someone opens the box and looks inside, collapsing the cat's so-called wave function into one state or the other. Though seemingly ridiculous at the macroscopic level of a cat in a box, this phenomenon holds at the level of fundamental particles, and led Einstein, along with Boris Podolsky and Nathan Rosen, to develop their own thought experiment. Take a radioactive atom and let it decay, sending two decay fragments off in different directions. According to quantum mechanics, they could have a number of properties, but until we look, we won't know what those properties are. That would mean that by observing one, collapsing its wave function, we would simultaneously know the properties of the other. Einstein called this 'spooky action at a distance', and used it to argue against quantum mechanics, since it would allow information to be transmitted instantaneously from one point to another, breaking the cosmic speed limit stipulated by the theory of relativity. Modern experimental techniques, however, have allowed it to be tested. Spooky action at a distance is real. It's counterintuitive, but fully consistent with theory.

All of these things made quantum mechanics revolutionary, but perhaps most significantly of all, quantum mechanics steered physics away from the deterministic route, saying instead that much of what happens is essentially statistical – a set of probabilities rather than absolute certainties. From now on, modern physics research would acquire a thirst for data: just because you saw a particular kind of particle decaying in a particular way once, for example, didn't mean that such particles would behave that way all the time. To get the full picture, you'd have to record a large number of decays.

Quantum mechanics and relativity are the two pillars of 20th-century physics, and while some perceive them as rather esoteric pieces of work with little practical use, we rely on them every day. Modern electronics is founded on quantum mechanics, making those GPS devices we take for granted nothing less than quantum relativistic machines. Although Einstein and Bohr were motivated by the quest for knowledge, not smartphones, it's usually through fundamental research such as theirs that the big technological advances arise.

Particle accelerators

While theoretical physics was undergoing a revolution, there were also developments in technology that would change the face of research, embodied by the devices called particle accelerators. Thomson had used a rudimentary accelerator when he discovered the electron back in 1897, although it was not yet called that. His device consisted of a tube from which the air had been removed, along with a particle source, and arrangements of electric and magnetic fields to accelerate and steer the particles. Combinations of these components are the basic ingredients of every particle accelerator since, from the Van de Graaff generator familiar to high school physics students as the machine that makes your hair stand on end to the mighty Large Hadron Collider at CERN.

Robert J. Van de Graaff invented his accelerator in 1929. For a while it proved to be a valuable research tool, but its basic technique of taking electric charge from one place to concentrate it in another until sufficient had been accumulated to generate a high enough voltage to usefully accelerate a beam of charged particles would ultimately prove impossible to scale up to the increasingly high particle energies that researchers needed for their experiments. Also in the 1920s, Norwegian accelerator pioneer Rolf Widerøe built a linear accelerator in which an alternating voltage is used to boost a particle beam's energy as it travels in a straight line. Linacs, as they came to be known, found applications in many domains and are still used in the early stages of many accelerator complexes today, including that of CERN. The world's largest linear accelerator today is at Stanford, California, and stretches for two miles. It has been the scene of many important discoveries.

Most large accelerators today, however, are not linear but circular. Their origins can also be traced to Widerøe, but those most familiar to particle physicists stem from Ernest Lawrence, who in the 1930s in California built a machine consisting of two D-shaped hollow electrodes facing each other to form a circle with a gap between their straight edges. A magnetic field was applied perpendicular to the electrodes, and particles were introduced at the centre of the circle. An oscillating electric field between the electrodes would accelerate the particles into one of them, where they would be bent in a circular path by the magnetic field, bringing them back to the gap to be accelerated across to the other electrode and so on. Each time they crossed the gap they would gain in energy and spiral outwards inside the electrodes. These machines are called cyclotrons, and despite the size limitation imposed by the need for them to be embedded in a magnet, they became the workhorses of particle acceleration. The largest in operation today is in Vancouver, Canada, at a laboratory called TRIUMF, where it provides beams for a range of research projects, and is also used to treat a certain form of cancer, ocular melanoma, for which traditional radiotherapy is not appropriate.

CERN's first accelerator would be a kind of cyclotron known as a synchrocyclotron. At low energies the revolution frequency in a cyclotron, and thus the frequency of the oscillating electric field, is constant with increasing energy. However, when a particle's energy reaches a significant fraction of the speed of light, relativistic effects come into play, and different rules apply: a big increase in energy equates to only a small increase in velocity. This means that the oscillating electric field accelerating particles across the gap has to be changed in synchronisation with the increasing energy of the accelerated particles. This principle allowed CERN's first machine to reach an energy of 600 MeV in 1957.

As we have seen, in particle accelerators, energy is measured in electronvolts, eV, which is the energy gained by an electron accelerated by a potential difference of one volt. An electronvolt is tiny, and even in the early days, a few million electronvolts, or mega-electronvolt, MeV, was common. By the early days of CERN, thousands of millions of electronvolts, gigaelectronvolts or GeV, were possible, and today we are into the realm of millions of millions – teraelectronvolts, TeV. Because of Einstein's famous equation, $E=mc^2$, physicists often refer to the masses of particles in electronvolts as well. Strictly speaking this should be electronvolts divided by c^2, but in particle physics shorthand, the mass of an electron is often quoted as 511 keV, for example, and that of a proton about 1 GeV – though to be strictly correct, it ought to be GeV/c^2.

Today's big circular machines are called synchrotrons and were invented independently in the USSR by Vladimir Veksler and in the US by Edwin McMillan in the 1940s. Synchrotrons solve the scalability problem by moving to a fixed circular orbit and changing both the frequency of the accelerating field and the bending magnetic field of a ring of magnets in a synchronous way with increasing beam energy. Acceleration in a synchrotron is performed with devices called radio frequency, or RF, cavities. These produce an oscillating electric field along the direction of beam. By correctly timing the arrival of particles at the cavities, the particles get an accelerating kick each time they pass through. The magnetic field in the bending magnet is increased in parallel with the energy gain to keep the beam on its fixed orbit. CERN's 27km Large Hadron Collider is the world's largest and most powerful synchrotron, designed to accelerate particles from an initial 450 GeV up to 7 TeV.

From atoms to particles

Earlier in this chapter we traced the development of fundamental physics from ancient Greece to the first half of the 20th century. By this time, we knew that atoms were made of protons, neutrons and electrons, and that the interaction between charges was what caused negatively charged electrons to join positively charged nuclei to form atoms. But we also knew that many more particles existed than were needed to make up ordinary matter. They had been observed in cosmic rays or the radioactive decay of certain elements. How they fitted into the overall picture remained a mystery.

We knew that terrestrial and celestial gravity were the same, and that electricity and magnetism were also manifestations of the same underlying physics. We knew, thanks to Maxwell, that light and electric charges interacted.

From the discovery of the electron to the 1940s, there had been an explosion of creativity in physics, both theoretical and experimental. On the side of theory, Einstein had refined our understanding of gravity with his theory of general relativity, while the pioneers of quantum mechanics had given us the theoretical underpinning for the interactions between particles. Einstein had given us a theory of the very large, while quantum mechanics provided the theory of the very small. Together, they covered everything, but they remained separate theories. Bringing them together would prove to be the ultimate challenge for those wishing to develop a single unified theory describing all the forces of nature.

For experimentalists, while studies of cosmic rays using high altitude balloons or observatories perched on mountaintops were still the main tool for investigations, particle accelerators offered a promising new avenue. Accelerators allowed scientists to make artificial cosmic radiation in the laboratory where they could determine the experimental conditions, rather than simply observing passively what nature provided. As the world emerged from the ravages of the Second World War, fundamental physics was full of questions to be answered, and poised on the cusp of a golden age of discovery.

FROM THE ASHES 3

European Movement

'The profoundly contradictory condition in which Europe has lived for ten years has entered the critical phase. It is almost desperate.' These were the words with which Swiss writer and European federalist Denis de Rougemont opened the European Cultural Conference in Lausanne on 8 December 1949. 'It is true that Europe is being undone,' he continued. 'She has never been more threatened, more divided in the face of danger, more anxious and sceptical at the same time. But it is no less true that for the first time, in its long history – consciously – Europe is in the process of building itself!'

De Rougemont was a leading member of the nascent European Movement that had emerged after the Second World War: a movement that had already scored a number of successes in promoting European cooperation. Earlier in 1949, the Treaty of London had established the Council of Europe to uphold human rights, the rule of law, and democracy across the continent, and the College of Europe in Bruges had been set up to provide postgraduate education to Europe's future leaders.

This was the construction that de Rougemont was referring to in his opening address, but he also went on to examine how culture could be leveraged to draw the formerly warring nations of the continent closer together, contributing to the ideal of making future conflict unimaginable. To de Rougemont, science was very much a part of European culture, and it had a vital role to play.

All this came against a backdrop of tension and global unease. 1949 had seen the establishment of NATO and the emergence of the People's Republic of China. The year had opened with the Berlin blockade in full swing, and by the time it ended, the Soviet Union would have entered the nuclear age. It was the year in which George Orwell captured the global mood with the publication of his chilling future dystopia in the novel *Nineteen Eighty-Four*. The European cultural conference offered a glimmer of hope for a brighter future than the omens were suggesting.

Something of a polymath, de Rougemont had had the opportunity to meet Einstein at Princeton in 1947, where they had discussed the idea of linking European unity with the control of nuclear energy. Once back in Europe, de Rougemont had gone to see Raoul Dautry, a former minister in the French government and the General Administrator of the country's Atomic Energy Commission, which operated accelerator- and reactor-based nuclear research facilities near Paris. Thanks to these facilities, France was emerging as the powerhouse of nuclear physics in continental Europe. As a consequence of his meeting with Dautry, de Rougemont invited the French Nobel Prize winner, Louis de Broglie, a significant contributor to quantum physics, to speak at the Lausanne conference.

De Broglie was not able to attend in person, and Dautry delivered his address for him. 'It is not only economically or politically that these movements seem desirable or even necessary; it is also on the intellectual plane, and especially on the scientific level,' Dautry said. 'At a time when we are talking about the union of the peoples of Europe, the question now arises of developing this new international unit, a laboratory or institution where it would be possible to work scientifically, in a manner outside and above the framework of the different participating nations. As a result of the cooperation of a large number of European States, this body could be endowed with more resources than those available to national laboratories and could subsequently undertake tasks, which by virtue of their size and cost, remain prohibitive to these. It would serve to co-ordinate research and results obtained, to compare methods, to adopt and to carry out programs of work, with the collaboration of scientists from various nations.'

Another to shape the landscape of European cooperation in particle physics was Lew Kowarski. Born in Russia in 1907, Kowarski moved to France via Poland and Belgium after the Bolshevik revolution, and was a leading contributor to nuclear science. As part of his work, Kowarski and his team acquired the world's entire stock of heavy water from Norway on the eve of its invasion, and later, when France was overrun, took it to England. Heavy water, in which the protons of the hydrogen nuclei are replaced by deuterium, a proton and a neutron, was deemed vital to nuclear weapons research at the time. When the war was over, a film, *Operation Swallow: The Battle for Heavy Water*, was made of these exploits in which Kowarski played himself, alongside Dautry and the renowned scientists Hans von Halban and Frédéric Joliot who had also been involved in the adventure. Not only would Kowarski become a founding father of CERN, he was also a genuine war hero. When the UN established the Atomic Energy Commission in the mid-1940s, bringing scientists and diplomats together, he summed up this somewhat unusual alliance. 'It was a pleasure to watch the diplomats grapple with the difference between a cyclotron and a plutonium atom,' he said. 'We had to compensate by learning how to tell a subcommittee from a working party, and how – in the heat of a discussion – to address people by their titles rather than their names. Each side began to understand the other's problems and techniques; a mutual respect grew in place of the traditional mistrust between eggheaded pedants and pettifogging hair-splitters.'

One particularly influential example of that mutual respect blossomed between the French diplomat François de Rose and the American scientist Robert J. Oppenheimer, leading to a further precursor for a European science project. As they discussed the state of science, the idea of promoting European research in nuclear physics began to take shape. Back in the 1940s, the term nuclear physics encompassed everything from applied research into nuclear energy to the very esoteric studies of cosmic rays, increasingly complemented by particle accelerators, to probe the structure of matter. Countries where these fields were well developed, like Britain, France and the USA, operated research facilities that ran accelerators and reactors on the same campus to explore the fundamental physics of the atomic nucleus. Could the nations of Europe combine forces to build a world-class laboratory covering the same ground?

Gaining momentum

The idea of creating a European laboratory was gaining momentum. By the time of the 1949 European Cultural Conference, it was ready to emerge. Although de Broglie's message did not specify any particular domain, Dautry proposed that astronomy and astrophysics, or atomic energy would be fertile ground for European collaboration, saying that, 'what each European nation is unable to do alone, a united Europe can do, and I have no doubt, would do brilliantly.'

The geopolitics of the time were complex to say the least, with any mention of atomic or nuclear research raising alarm bells. The Manhattan Project and its consequences in the atomic bombs dropped on Hiroshima and Nagasaki were fresh in people's memories, and while the post-war world order was becoming established, countries with a capacity in nuclear science kept it to themselves. The idea that a fragile post-war Europe, potentially susceptible to the siren call of communism, could collaborate with the USA was unthinkable to everyone. And in Europe itself, Britain was perceived to be sufficiently autonomous that the view from many on the continent was that an international collaboration between continental European countries, without the UK, would be necessary for the science to advance there. Would global politics allow that to happen? The answer would be some time in coming.

Many of the delegates at the Lausanne conference were scientific administrators like Dautry, fervent Europeans who saw a peaceful future through increasing cooperation and integration. It's perhaps not surprising, therefore, that they grasped the nuclear bull by the horns and advocated the foundation of an institute for nuclear physics in its application to daily life. Such an institute would be a triumph for the

European Movement, showing that political divisions could be overcome even to the extent of collaborating on nuclear energy. It was a bold vision, and one that turned out to be very different from what would eventually emerge as CERN.

Following the Lausanne conference, France, and in particular Lew Kowarski, took a leading role. Kowarski was the first to suggest that the proposed institute focus on pure research only. It was clear to everyone that an intergovernmental research organisation could have nothing to do with military research, but Kowarski was the first to suggest that applications in energy should also be left to others. He proposed setting up a laboratory for fundamental research with a relatively modest accelerator of 500 MeV–1 GeV and a research reactor. He also proposed a budget of 5–10 billion French francs per year, modest in comparison to the equivalent of 30 billion francs invested by the UK and 300 billion by the USA. The governance structure he proposed was similar to that in place at CERN today, with a governing body made up of government representatives, an executive structure to manage the laboratory and a council of scientific experts to recommend the research programme.

Kowarski's ideas attracted the attention of the French government, and of people like François de Rose, but in the 1940s, these ideas were going nowhere. Perhaps it was because national research budgets were already too tight to consider establishing an international laboratory. Or perhaps the involvement of a research reactor took the proposal too close, in people's perceptions, to military or industrial research for comfort. Whatever the reason, the idea seemed to have run out of steam. But then, in June 1950, everything changed with the arrival of a new protagonist for a European laboratory, this time from the other side of the Atlantic.

American influence

Isidor Rabi had been awarded the Nobel Prize in physics in 1944 for 'his resonance method for recording the magnetic properties of atomic nuclei'. This was to become known as nuclear magnetic resonance, NMR, and has spawned many applications today, including the powerful medical diagnostic technique of magnetic resonance imaging, MRI, which gives detailed pictures of body tissues without using X-rays.

In June 1950, Rabi was an American delegate to the fifth General Conference of UNESCO in Florence. By this time responsibility for exploring the possibility of establishing international labs had passed from the UN to UNESCO, and Rabi was surprised to see that there was nothing on the agenda on that subject. He tabled a resolution calling on UNESCO to work towards establishing regional research centres. In presenting the resolution, Rabi stressed that his initiative was primarily intended to help countries that had previously made great contributions to science, and he referred to the creation of a centre in Europe. One of the factors that had spurred the European scientists and science administrators from the start was the brain drain from Europe, in particular continental Europe, during the war. A major centre on European soil would help restore balance to the transatlantic science landscape.

Rabi also made it clear that he was acutely aware that in certain fields of research, the US and perhaps the UK held a virtual monopoly, and that the war had forced European countries to lose their leading roles. A consensus was emerging that healthy science in Europe equated to healthy science on both sides of the Atlantic. Rabi had been a major proponent of the Brookhaven laboratory, established in 1947 at Camp Upton, a former military training establishment near the town of Brookhaven on Long Island. A focal point for US nuclear physics, Brookhaven would be run by a consortium of universities, and have both reactor and accelerator facilities. Rabi's proposal was for a European counterpart, but with just an accelerator, and with countries as partners rather than universities. Having no reactor would make the politics easier, and would allow West Germany to take part: after the war, West German scientists were expressly forbidden from working on reactors.

There were other political undertones to Rabi's resolution. The explosion of an atomic bomb by the USSR on 29 August 1949 had underlined the Soviets' nuclear capabilities, and it was clear that transatlantic cooperation would be needed as the nascent cold war set in. Furthermore, the Americans had recognised that the US had benefited perhaps more than any other world power from scientific discoveries made elsewhere, and that American scientific strength lay in the application of such discoveries. With US science again focusing on military research, it would be convenient for the US to have a source of openly accessible fundamental physics results coming from Europe, for US laboratories to apply.

The full reasons for American support have been debated at length, but the net result was that, coming as it did from an American with explicit approval from Washington, Rabi's resolution gave new life to the European initiative. Resolution No. 2.21 to set up 'regional research centres and laboratories' was unanimously approved, and the director of the UNESCO natural sciences department, French physicist Pierre Auger now had a mandate.

Getting off the ground

The rest of the year was punctuated by international meetings and discussions. At one point, the name of Niels Bohr, pioneer of quantum mechanics and since 1921 head of the eponymous Copenhagen institute that had become a world-leading centre for theoretical physics, was brought into the discussion. He was reported to have expressed the opinion that the construction of a reactor would bring in a range of complex economic and political questions that might prevent the project from ever getting off the ground. In December 1950, Pierre Auger told a meeting at the European Cultural Centre in Geneva that he would be focusing on a laboratory based around a high-energy particle accelerator. Although the term particle physics did not yet exist, today's CERN was starting to emerge.

The Geneva meeting was also significant for another reason: it was the first time that money was committed to the project. With Italy's delegation pledging funds at the meeting itself, and France and Belgium following soon after, the princely sum of $10,000 was put together. It wasn't much, but along with UNESCO's support, it was enough to get the project off the ground.

Auger lost no time. He created a special office at UNESCO to coordinate the project, and personally co-opted the brightest and the best in European physics as advisors, starting with Italian physicist Edoardo Amaldi. In May, Auger and Amaldi were ready to call the first meeting of their board of consultants at UNESCO's headquarters in Paris. There were representatives from many European countries including Kowarski from France, Odd Dahl from Norway and Frank Goward from the UK, a sign that the UK might perhaps join the continental endeavour after all. In total, the board consisted of eight people.

Two initiatives emerged from that meeting: the ambitious plan to build a particle accelerator that would be second to none, along with the more modest aim of building a less powerful machine that could rapidly be up and running to keep momentum building. On learning the news, the British physicist, Herbert Skinner from Liverpool University, went on record to say that the plan to build the world's biggest accelerator was among the 'high-flown and crazy ideas which emanate from UNESCO'. Fortunately for European science, he turned out to be on the wrong side of history, and would later open the doors of his laboratory to help the European endeavour get off the ground.

The majority of consultants put their weight behind the bolder initiative, and went on to propose that it be developed in stages rather than trying to persuade governments to commit at a single stroke. Phase one would be a provisional organisation tasked with producing a plan and budget for the first six or seven years of the proposed laboratory. To do so, the provisional organisation would have twelve to eighteen months and a modest budget in the region of $200,000. A decision was taken in May 1951 that would become part of CERN's ethos. The consultants decided that everything produced by the provisional organisation would be made freely and openly available. Almost 70 years later, CERN is still in the vanguard of the open access movement, advocating for open access to scientific publications, promoting open approaches to software and hardware, and opening its doors to over 100,000 public visitors each year.

The consultants met twice more in 1951, and by the end of the year UNESCO was ready to call an intergovernmental meeting of interested countries. A letter of invitation was sent to all of UNESCO's European members, but when François de Rose opened the gathering in Paris on 17 December, the chairs reserved for countries from the east of the continent remained empty, with the exception of Yugoslavia's. From this point on, the laboratory would be developed as a largely western European project. A total of 21 countries were present, and debate was intense. The British Delegation, led by George Thomson, son of the discoverer of the electron and himself a Nobel laureate, argued in favour of building collaboration around existing facilities rather than embarking on an ambitious international project. He argued that people were more important than machines, and offered the soon-to-be completed 400 MeV Liverpool cyclotron as an example of a facility that could be made available to scientists from all over Europe. The notion was not without support from the cash-strapped countries of Europe, with Werner Heisenberg notably putting his weight behind Thomson. An alternative view, voiced by the Yugoslav delegate, was that people would go to where the most powerful machines are.

A second intergovernmental gathering in Geneva in February 1952 resulted in an agreement to constitute a 'Council of Representatives of European States for planning an International Laboratory and organising other forms of co-operation in Nuclear research'. At the meeting's conclusion, a letter was rapidly dispatched to Rabi at Columbia University in New York saying: 'We have just signed an Agreement which constitutes the official birth of the project you fathered at Florence. Mother and child are doing well, and the Doctors send you their greetings.' It carried eleven signatures. Geneva was chosen as the Council's seat, and over the following three months the agreement was ratified, coming into force on 2 May. Eleven countries signed the agreement, but the UK preferred to remain an observer. That did not stop the British delegates from playing an active part, however, and even making financial contributions to the fledgling initiative. The agreement was limited to eighteen months, at the end of which the Council was supposed to have produced a convention for the establishment of a new laboratory. More hints of the way CERN would be organised were visible in the agreement; each member country would have two seats on the Council, for example, and decisions would be taken on the basis of one vote per country.

CERN

On the minutes from the Council's first meeting, the rather cumbersome name had been replaced by the somewhat more compact European Council for Nuclear Research, or Conseil Européen pour la Recherche Nucléaire, accompanied by the acronym 'CERN'. From here on, a series of meetings hammered out the shape of the nascent CERN's development. The laboratory's first machine was planned to be a synchrocyclotron with an energy of at least 500 MeV. By the second meeting, a report drafted by Werner Heisenberg concluded that over the last two decades, the centre of interest in atomic physics had changed from the nucleus to the elementary particles, saying that in order to understand nuclei, it would first be necessary to understand the elementary particles. They went on to say that to do that, particle accelerators would be essential because, although unable to match the energy of cosmic rays, accelerators provided much more intense beams, giving physicists more particle interactions to study.

Heisenberg and his colleagues noted that there were no

accelerators in Europe above 200 MeV, although some were nearing completion in Britain, while in the USA, a 3 GeV machine called the Cosmotron was already in operation at Brookhaven, on Long Island in New York state. The success of the Cosmotron, coupled with the state of physics in 1952, led the protagonists for the European laboratory to conclude that: 'if one wants to break entirely new ground, one has therefore to consider the construction of a big machine in the 10 GeV region.' The goals of the new laboratory were taking shape.

A team visited the Berkeley and Brookhaven laboratories in the US, home of the largest and most powerful accelerators in the world. At the Long Island laboratory, they learned some very exciting news. Three Brookhaven scientists, Ernest Courant, M. Stanley Livingston, and Hartland Snyder had developed the new concept of alternating gradient, or strong-focusing. In this scheme, the accelerator's quadrupole focusing magnets are arranged to be alternatively focusing and defocusing. A quadrupole that focuses in the horizontal plane defocuses in the vertical, and with the proper arrangement of magnets, the overall effect is strong focusing in both planes. This opened the way for higher beam intensity and smaller vacuum chambers, and thus reduced costs, allowing the investment to be made in a larger machine that could reach higher energy. The Brookhaven scientists were generous in sharing their new ideas, launching an era of trans-Atlantic cooperation in particle physics that remains a hallmark of the field today.

At the third meeting of the Council in Amsterdam in October, Odd Dahl presented two proposals for the Proton Synchrotron, Project 1 being the original 10 GeV machine, and Project 2 a 30 GeV machine using Brookhaven's alternating gradient technology, which could be built 'faster and cheaper' than the Project 1 machine. The Council chose Project 2. In terms of sheer ambition, it was the LHC of its day. The energy would later be reduced to 25 GeV to keep the cost under control without unduly limiting the research potential.

At the same time, Geneva was chosen as the home for the future laboratory, selected from proposals submitted by the Danish, Dutch, French and Swiss governments. While all were deemed to be good candidates, Geneva was selected by a unanimous vote. Its central location in Europe, Swiss neutrality during the war and the fact that Geneva already hosted a number of international organisations all played a role. While preparations were being made to establish the laboratory in Geneva, theoretical work would be carried out in Copenhagen.

By the sixth meeting in Paris on 29–30 June, the provisional CERN had done its job, within the time allotted. The draft Convention was complete and approved unanimously by the representatives of the eleven countries that had signed the original agreement plus the UK, and the document was made available for signature.

The CERN Convention is a remarkable document. It is very concise, running to just over a dozen pages, but it is a blueprint for harmonious international collaboration. According to the Convention, financial contributions are calculated on the basis of net national income averaged over recent years so that each member state pays according to its means, though there is a cap to ensure that no single country ever pays more than a maximum percentage of the overall budget fixed by the Council, and there are provisions to help member states experiencing financial difficulties. In short, those who drafted the CERN Convention showed great foresight, setting the organisation up for a long and productive future. They had done a great job, and all that remained now was to collect the signatures.

A NEW LABORATORY IS BORN

4

On 29 September 1954, Frank Sinatra topped the UK singles chart with 'Three Coins in the Fountain'. William Golding's *Lord of the Flies* had just been published, while earlier in the year, post-war food rationing had ended in Britain and West Germany had won the football World Cup, beating Hungary 3-2 in the final. On that day in September, France and Germany deposited their instruments of ratification of the CERN Convention at UNESCO House in Paris, formally bringing the European Organization for Nuclear Research into existence. At this point, the interim Council ceased to exist, and the acronym CERN should also have been consigned to history, to be replaced by OERN. By this time, however, the name CERN had stuck, and it remains in use to this day, even though the body it represents was wound down over 60 years ago.

Before that momentous day, CERN had existed in a kind of limbo. With the approval of the CERN Convention in June 1953 at the Council's sixth meeting in Paris, the provisional CERN had been wound up, giving way to an interim

CERN, whose job was to pave the way for the establishment of the European Organization for Nuclear Research, putting in place rules and regulations and drawing up the legal agreement between the CERN-to-be and its host state, Switzerland. In short, the interim CERN was to set up all the administration and plans needed to launch and operate a new intergovernmental organisation. Nothing more.

The mood in Paris had been optimistic and nobody on the Council had expected the ratification process to take long; a six-month timescale had been agreed: up to the end of the year. As a temporary organisation, the interim Council was still operating under the auspices of UNESCO and should not have undertaken any tangible actions until CERN became a legal entity in its own right. What actually happened was rather different. The Council members, now chaired by the Frenchman Robert Valeur, wasted no time in preparing to establish a laboratory in Geneva. Odd Dahl's proton synchrotron study group was installed at the University of Geneva's physics institute, and an embryonic directorate and administrative group got to work, housed in the Villa Cointrin at Geneva airport. Under Valeur, the Council held three sessions, all in Geneva, before the Convention was fully ratified.

To Valeur's Council, high on the euphoria of June 1953, there seemed to be no time to lose, but they hadn't bargained for the time it would take for the Convention to pass through the democratic processes of each founding CERN member state. By the end of the year, only Britain had ratified, depositing its instrument of ratification at UNESCO House on 30 December. Switzerland was second to ratify in February 1954, followed by Denmark in April, the Netherlands in June, and Greece, Sweden and Belgium in July. When France and

Germany followed on 29 September 1954, all the conditions for bringing the new organisation into existence had been fulfilled: at least seven member states had ratified, including the host state Switzerland, and between them they represented over 75 per cent of the organisation's overall budget. The process concluded with Norway ratifying in October 1954, and Yugoslavia and Italy following in February 1955.

While ratification was winding its way through the various parliaments, Valeur chaired three interim Council meetings, the first in October 1953. Bakker reported that the relatively small synchrocyclotron (SC) would be operational within just three to four years of CERN formally coming into existence. There was much discussion about the energy, cost and timescale of the proton synchrotron (PS). Some delegates argued that it would be better to have a 20 GeV machine ready as soon as possible rather than a more powerful machine later. Others argued that the physics potential might be limited at the lower energy. While a machine designed to reach 20 GeV could comfortably study meson production, it might struggle with nucleon-antinucleon production, which was seen as a potentially rich source of new physics: at the time of these discussions, antinucleons were still hypothetical particles.

British physicist John Cockcroft broke the impasse, proposing a machine that could achieve 20 GeV with ease, but that had the possibility of being pushed to 25 GeV. In his view, such a machine would have a nucleon-antinucleon pair production rate high enough for meaningful scientific study. The meeting settled on 25 GeV, and Dahl's team got to work.

It was also at Valeur's first meeting that delegates thrashed out the framework for a governing structure. A Director General appointed by the Council would run the laboratory, supported by a number of divisional directors. Broadly, it is a structure that remains in place. Valeur's first meeting in the chair was a productive one, but still not a single member state had ratified.

During Valeur's term of office, the Zurich-based architect Rudolf Steiger, a proponent of the Neues Bauen School of architecture that had remodelled much of Europe before the war with its clean, striking lines, was chosen to produce a bold new design for the laboratory. Because of the lack of ratification, a termination clause was proposed to the architect in case the whole thing fell through. He agreed, and the earth-moving machines got to work on 17 May 1954. It was a historic moment, but few were there to see it other than a handful of CERN staff along with Robert Valeur and some Geneva officials.

By this time, a Director General had been identified, Swiss-American physicist Félix Bloch. A Nobel Prize-winning professor at Stanford University, Bloch's expertise was nuclear magnetic resonance – not a field directly related to CERN's research but Bloch was nevertheless persuaded to accept the job on the basis that someone of his stature was needed in the role. As Valeur's final meeting in the chair closed in April, Valeur expressed the hope that it would be the last meeting of the interim CERN, a wish that would be fulfilled.

By the time the Convention was finally ratified, CERN was firmly in the starting blocks. Committee structures were in place, major contracts were ready to be signed, and some 120 members of staff were already at work, mostly in Geneva, with some in Copenhagen and a handful in Liverpool. An architect had been chosen, and construction work on the site had begun. Valeur's time in the hot seat had been remarkably fruitful.

Architectural developments

To the thousands of commuters who cross the border from France to Switzerland to work every day, CERN above ground is hardly an architectural gem. It is true that the flags of the member states, accompanied by those of the Canton of Geneva and CERN itself, and the striking Globe of Science and Innovation donated by the Swiss Confederation to CERN in 2004 to house a visitor centre, hint at the importance of what goes on here, but investment in architecture, at least above ground, is not the top priority for CERN. It has not always been so. When Steiger was selected to design and build the laboratory in 1954, he provided a statement of functional modernity, or rather his son Peter did. Steiger senior was busy on the new Cantonal Hospital in Zurich, so Peter, just 21 at the time, was handed the project. He was young and inexperienced, but he'd spent time in the US working with Frank Lloyd-Wright, and to aficionados there are hints of the great architect's influence to be seen at CERN. Peter Steiger's youth did not dissuade him. Instead, the chance to manage such a large project spurred him on. As he pointed out in a 2010 interview, 'I didn't yet have the experience of such a large project, but by chance everyone at CERN was young and inexperienced!' He found himself in good company.

Steiger delivered a coherent complex of administrative buildings, workshops and laboratories, and of course the buildings to house the SC and PS accelerators. The focal point was the so-called CERN Main Building, with its bold and elegant lines, a grand entrance with monumental staircase, tiled mosaic floor and supporting pillars reminiscent of Lloyd-Wright's famous inverted mushroom columns. Even outside, the street lighting was elegant, with pairs of curved lampposts delicately intertwining like the necks of courting swans floating on a summer lake. Today, the original CERN campus is hard to see, crowded out by newer and less remarkable buildings, but the Main Building remains the heart of the laboratory. Asked his opinion on the changes wrought to his grand design, Steiger was stoical. 'Buildings from the 50s and 60s aren't yet recognised as heritage,' he said, 'they are still very young. Future generations will take an interest, but we're not there yet.'

Things were progressing well, but it was not all plain sailing. A popular initiative spearheaded by the local Community Party to block the establishment of CERN in Geneva received enough signatures to force a referendum in the Canton, and attracted support from those who felt that such a laboratory would compromise Switzerland's famous and prized neutrality. The question went to the vote on 28 June 1953, a month after work had begun on the site, and was roundly defeated with some 70 per cent of votes cast in favour of the laboratory.

Down to business

Robert Valeur's final duty as Chair was to open the first session of the CERN Council proper, in Geneva on 7 October 1954. Edoardo Amaldi was there as Assistant Director General, as was Félix Bloch as head of the new laboratory. Business as usual was rapidly established as the agenda moved on to reports from the working groups. Dahl and a member of his group, John Adams, introduced what would be a recurring theme throughout CERN's history: the need for extraordinary precision in civil engineering. The PS would be a circle of 100m radius with components aligned with an accuracy better than 1mm. Requirements such as these dictated the extent of the building's foundations, which would have to be extremely stable, and led the Council towards a discussion of the geology of the Geneva basin, a region that had been under hundreds of metres of ice during the last ice age.

Soon after that first meeting, the question of what the laboratory should be called came up in a memo from the Director of Administration, Samuel ffrench Dakin. The memo was entitled: 'Conflict between Title and initials of the Organization'. That might seem like a trivial detail, but it had significant procedural and legal consequences. Dakin argued that there was no reason why the European Organization for Nuclear Research should not be nicknamed CERN, as long as it was explicitly pointed out in every legal document that the full title and CERN were interchangeable. But he also pointed out that Kowarski found the idea 'so silly as to be intolerable'. Dakin drafted a letter for Bloch to send to member state delegations asking for their opinions, pointing out that OERN is difficult to pronounce in most languages, and noting that any change would require the lengthy procedure of the Council making a recommendation that would have to be accepted in writing by all member states. It seems that no one felt that strongly about the issue, so CERN remained. Discussions about the name of the organisation continue to this day, though now the focus is more on why the word 'nuclear' appears in the name of a laboratory whose main line of research is particle physics.

On day one of the second meeting of the CERN Council in February 1955, Italy deposited its instrument of ratification with UNESCO, bringing the ratification process to a close. The biggest news from this meeting, however, was that Bloch was leaving. He announced his wish to resign because the onerous nature of the administrative duties allowed him too little time for research. The Council accepted his resignation with regret, and unanimously appointed Cornelis Bakker to take over on 1 September 1955.

Among Bloch's remaining duties as Director General was the official laying of the foundation stone. This was to be done in the presence of Max Petitpierre, President of the Swiss Confederation. On 10 July 1955, Bloch duly placed a small document written in English and French, the official languages of CERN, inside a steel canister and sealed it in the foundations of the building that would house the SC. The document read: 'On this tenth day of June, one thousand nine hundred and fifty five, on ground generously given by the Republic and Canton of Geneva, was laid the foundation stone of the buildings and headquarters of the European Organization for Nuclear Research, the first European institution devoted to co-operative research for the advancement of pure science.'

Things were progressing well at the Geneva laboratory, and there were also advances in physics elsewhere. In 1956 in the USA, Frederick Reines and Clyde Cowan finally identified the particles that Wolfgang Pauli had proposed in his open letter of 1930. The Europeans were raring to go, and it would not be long before research could get under way at the first of CERN's accelerators. Bakker was good to his word: the SC was operational within three years of day zero, on 1 August 1957. The logbook entry for that day reads: 'The 1st of August 1957 we had a short celebration after the appearance of the first circulating beam.' Massive for its time, the synchrocyclotron was comfortably housed in a single building, and it went on to set a proud CERN tradition of longevity, running for 33 years before being decommissioned in 1990. Today, the SC is still hard at work in a different way: as a visitor attraction recounting the early history of CERN.

THE BIRTH OF THE BIG MACHINES 5

'Remember the night of 24th November 1959? Of course I do.' So began Hildred Blewett in her wonderful reminiscences of the night that the proton synchrotron (PS) became the highest-energy particle accelerator in the world. A Brookhaven scientist, she and her husband John had been part of the PS group during CERN's early days, and she was back for the start-up – or so she hoped. Time, though, was running out. Her passage back home to New York was booked, and she'd be leaving Geneva on 25 November. With less than 24 hours to go, as she and project leader John Adams ate their dinner in the CERN canteen, her hopes of seeing the machine in action before she left were not high.

It had been an eventful five years since CERN had formally come into existence. At first, it was hard to generate an intellectual scientific life at the laboratory: without any operating machines, there was not much for an experimental physicist to do, and CERN had difficulty attracting researchers. That all changed in 1957 when the first beam circulated in the synchrocyclotron (SC) at its design energy of 600 MeV – making the SC the highest-energy particle accelerator in Europe, though still some way behind the big multi-GeV machines in the US.

Among the particles that had been discovered in the first half of the 20th century were pions, which were unstable. A negative pion, for example, would decay into a muon and a neutrino, which in turn would decay into a stable electron, along with a further pair of neutrinos. By the time the SC started up, there were compelling reasons arising from theories of particle interactions to think that occasionally, but very rarely, the decay process could skip the muon, with the pion decaying directly into an electron. Giuseppe Fidecaro, an Italian physicist who had first worked for CERN in Liverpool, and his colleagues realised that they could test this at the SC with a relatively simple experimental set-up. They would capture pions in a material called a scintillator, which produced a signal when a pion was stopped, and use a further array of scintillators to look at the particles emerging when the pion decayed.

When the team ran the experiment in 1958, most of the time they saw signals corresponding to a pion, a muon and an electron, but occasionally, roughly one time in ten thousand, the muon was missing, indicating that the pion had decayed directly into an electron, just as the emerging theory had predicted. This was the first of CERN's great discoveries, and as Fidecaro later said, 'The news went all over the world overnight'. The announcement was made in September at the UN's Atoms for Peace Conference in Geneva, and some 80 newspapers and magazines carried the story. CERN's fame was growing in other ways too, people from around the world wanted to visit this new research icon, and the lab's public information office found itself showing over 6,000 people from 58 countries around the premises over the course of the year.

1958 was also the year when work got under way in earnest on more sophisticated experimental apparatus. CERN's management had taken the decision that the proton synchrotron should be equipped with state-of-the-art particle detection systems to observe the interactions of the high-energy PS beams with fixed targets, allowing the particle interactions to be observed and measured. Today, particle detectors are usually electronic devices, but in the 1950s, they were essentially optical, with devices known as bubble chambers dominating the field.

Bubble chambers consisted of a liquid target held under pressure just below boiling point. When high-energy charged particles passed through the liquid, the pressure was reduced, allowing traversing charged particles to trigger the formation of bubbles. As the charged particles ionised the liquid, liberating electrons from atoms, the bubbles would form along the paths of ionisation, leaving trails of tiny bubbles like beads on a thread. These were photographed from several angles to allow the tracks of bubbles to be reconstructed in three dimensions for later analysis. Many important discoveries were made with bubble chambers, which remained in use right up to the 1980s, and at the peak of the bubble chamber era, there would be armies of people scanning the photographs one by one.

In 1958, CERN had a small 10cm liquid hydrogen bubble chamber and was constructing a larger 30cm chamber, but the laboratory had decided to build one at the unprecedented size of 2 metres. This would take time, and there was a chance it would not be ready before the PS started up. Fortunately, however, two other alternatives were on the table. The British were working on a 1.5-metre chamber destined for use at Harwell, but scheduled to be complete before it was needed there. The British chamber could be lent to CERN for the intervening period. The French were also building an 80cm chamber that they hoped to use at CERN. In addition to these liquid hydrogen chambers, CERN also took the decision to build a heavy liquid bubble chamber. A heavy liquid provides a high-density target without the need for cryogenic cooling. Though the results from such a chamber would be less clear than those from a liquid hydrogen chamber, the fact that it dispensed with the need for a cryogenic system to liquefy hydrogen meant that the operation of the apparatus would be less delicate and, crucially, that it could be completed more quickly.

Another highlight of 1958 was the arrival of CERN's first computer. The Ferranti Mercury was put to work straight away, among other things to help refine the pion decay measurements at the SC. The Mercury was used to run Monte Carlo simulations of detector elements used in the experiment. Monte Carlo techniques, so called because they are based on a random number generator – analogous with the roulette wheels for which that Mediterranean city is known – are an invaluable tool in particle physics. They are used to simulate what happens when particles interact with particle detectors, and for discoveries of new phenomena they play a vital role. As well as measuring what actually happens in the experiments, physicists generate a large quantity of Monte Carlo data to show what the experiment should see according to known physics. Any discrepancy between measurement and Monte Carlo can be evidence of something new.

Nescafé tin to the rescue

By 1959, the SC was an established part of the research landscape in the emerging field that now had the name of high-energy particle physics. The SC had put CERN on the map, but this was destined to be the year of the proton synchrotron.

On 22 May, the linac that provided beams for the PS accelerated its first particles to 10 MeV, a modest start. By the end of August, the linac was fully commissioned with a 50 MeV beam ready to be injected into the proton synchrotron. The PS had its first beam moment on 16 September when injected particles went all the way around the ring for the first time. This was an important milestone, and it came just in time for John Adams to announce the news at the second International Conference on High Energy Accelerators and Instrumentation, which was being hosted by CERN.

By October, they had managed to capture the beam – it is one thing to steer a 50 MeV beam around the ring using the magnets, quite another to time things such that the particles arrive at the accelerating radio frequency (RF) cavity at just the right moment to be marshalled into discrete and orderly bunches. When that happens, it's called beam capture, and it allows the beam to be stored, in principle, indefinitely. The next step is to accelerate the beam, and there were some successes in this respect through October and early November as beams reached energies of 2–3 GeV, but no more. Slow progress was expected with such a radically new machine, but even with these modest successes giving cause for optimism, everybody knew that the formidable barrier of transition still lay ahead.

There is a natural energy spread among the particles in a bunch, and this has two important implications. Firstly, higher-energy particles are deviated less by the bending magnets and travel further as they go around the ring. Secondly, the higher-energy particles move faster than the lower-energy ones. However, as the particles approach the speed of light, relativistic effects mean that the velocity gain with increasing energy is less and less, and eventually there comes a point at which all the particles take the same time to orbit the machine. This point is called transition.

Below transition, the higher-energy particles take less time to complete a lap than the lower-energy ones. The oscillating electric field in the RF cavities is timed such that the beam arrives as the field is rising so that the higher-energy particles see a lower electric field and gain less energy than the lower-energy ones that arrive later. This keeps the particles bunched and ensures a stable beam.

Above transition, the higher-energy particles take longer to orbit the ring than the lower-energy ones, so the RF system has to provide more energy to the lower-energy ones that arrive first. This means that the timing has to be flipped at the transition point so that the particles arrive as the electric field is falling, otherwise the beam becomes unstable. In the PS, transition happens at around 6 GeV.

Back in Geneva at Adams' invitation, Hildred Blewett was there to take part in getting the machine up and running so that she could learn from CERN's experience when Brookhaven started up its own big machine, the Alternating Gradient Synchrotron, AGS, the following year. 'I had to go back soon to help on the AGS,' she wrote. 'Pressure for high-energy protons in the United States was mounting even higher with the imminent production of European ones.' The weeks passed, and so she found herself in the CERN canteen with Adams on her last night in Europe. They finished their dinner and headed over to the PS. Since Adams' announcement at the conference, things had been moving slowly. The days were spent completing the installation and testing of equipment in the PS ring and control room, with only a couple of evenings a week devoted to beam tests.

As they walked towards the PS Central Building, Adams and Blewett were discussing an idea that the brilliant accelerator physicist, Wolfgang Schnell was working on. 'Wolfgang thinks this radial phase control will really work,' said Blewett. 'He's very optimistic, and maybe ...' She trailed off. Wolfgang may have been enthusiastic, but nobody else had any great hopes. Undaunted, Schnell had built his device to flip the phase of the accelerating field at the transition point using whatever he could find, including an old Nescafé tin, which happened to be the right size for one component. With little expectation, Adams and Blewett walked into the PS Central Building to find a sea of smiling faces. Schnell pulled them over to the oscilloscope monitoring the beam. 'We looked,' remembered Blewett. 'There was a broad green trace ... what's the timing ... why, why the beam is out to transition energy? I said it out loud – TRANSITION!' Schnell was soon back behind the electronics racks with his Nescafé tin, and before anyone knew it, the beam was through transition and up to 10 GeV. Later that evening, the beam was accelerated all the way to full energy. The PS logbook noted simply: '19:35 historic moment.' Blewett was less circumspect. 'It must be 25 GeV! I'm told I screamed, the first sound, but all I remember is laughing and crying and everyone there shouting at once, pumping each other's hands, clapping each other on the back while I was hugging them all. And the beam went on, pulse after pulse.'

When everyone came back down to earth, Blewett dashed

In the PS control room on the night of 24 November 1959. John Adams is on the left, accompanied by (left to right) Hans Geibel, Hildred Blewett, Chris Schmelzer, Lloyd Smith, Wolfgang Schnell and Pierre Germain.

CERN

off a telegram to Brookhaven to give them the good news and at 2am New York time, her phone rang. The news had spread rapidly across the US, and everyone wanted to know what had brought the success. When she told them about Schnell's Nescafé tin, there were smiles across the pond as well. Although not relying on a coffee tin, the Americans were building a similar device for the AGS, and were relieved to discover that that was what had done the trick. The AGS would come on stream in 1960, with a new world-record energy of 33 GeV.

Hildred Blewett was not the only American visitor at CERN in November 1959. Lloyd Smith, on leave from Berkeley, was also there for the historic moment. Blewett recalled how Smith went back to Berkeley with plans for a similar machine of 100 GeV, which would later evolve to a design for a 200 GeV machine. This was eventually built in the small town of Batavia, Illinois, on the mid-western prairies outside Chicago at the National Accelerator Laboratory, now better known as Fermilab, which was founded in 1967. Even back in the heady days of the 1950s, the tradition of sharing ideas in pursuit of common goals was alive and well in the field of high-energy particle physics.

Throughout the following year, the engineers gradually handed over more and more time to the experimental physicists so they could launch a research programme at the PS. Meanwhile, the SC continued to deliver first-class research, notably with an experiment designed to measure a property of the muon, known to physicists as g-2,* that was linked to one of the key questions of the time: did the theory of quantum electrodynamics, QED, which had been developed to describe the behaviour of electrons, also describe the behaviour of muons, and therefore establish itself as a theory for all kinds of electromagnetic interactions?

The muon seemed to be identical to the electron, apart from being roughly 200 times heavier and having a very short lifetime, and QED made very precise predictions for

* Muons act like small gyrating magnets, hence the letter g, which represents their magnetic strength. Classical theory predicts g to have the numerical value of 2 for muons, but quantum effects introduce a tiny difference, hence the interest in measuring g-2.

the muon's g-2. The experiment found a result that was very much in line with QED's predictions, underlining the strength of the theory for describing electromagnetic interactions. With the PS coming on stream, however, the SC's role at CERN's high-energy frontier was coming to a close, and CERN's first machine would soon be repurposed. From the mid-1960s right up to 1990, the SC's job was to supply an online isotope separator, ISOLDE, which could deliver beams of unstable isotopes covering the full span of the periodic table of the elements. ISOLDE is still in use today, though supplied from a different source, enabling experiments ranging from medical research to particle astrophysics.

As the 50s gave way to the 60s, CERN was coming to life in other ways. From May 1959, the CERN Council was able to hold its meetings at the laboratory, while in June, the Director General and his administration moved into the new CERN Main Building. CERN's early successes also came with other advantages: a new member state, Austria, was formally admitted to the organisation on 1 July 1959, and by the end of 1960, the number of staff and visiting scientists had grown to over 1,000.

With Cornelis Bakker having been at the helm through the triumphant start-up of the proton synchrotron, the CERN Council decided to reconfirm him in post for a further term of five years – but it was not to be. Bakker died tragically in a plane crash in the US on 23 April 1960. John Adams briefly stepped into the breach, but he had recently accepted the job of leading the development of the new Culham research centre in Oxfordshire, England and in August 1961, Victor – Vikki – Weisskopf became CERN's fourth Director General.

Physics at the beginning of the 1960s

The PS in Europe and the AGS in the US came on stream at a time of great opportunity for the field of high-energy particle physics. Barely half a century after the discovery of the first fundamental particle, physicists had made huge advances in understanding the underlying structure and behaviour of the universe's basic constituents. The discovery of the electron, followed by the atomic nucleus with its constituent protons and neutrons, seemed to offer a simple explanation of the constituents of matter. Subsequent particle discoveries, such as the neutrino and anti-particles fit comfortably into this picture. The development of quantum mechanics provided a theoretical framework to account for the interactions that held the basic constituents together to form complex objects.

Photons carry the electromagnetic interaction, which holds electrons in orbit around nuclei, and a similar particle, originally called the mesotron, but soon shortened to meson, was proposed in 1934 by the Japanese theorist Hideki Yukawa to carry the strong interaction that holds the nucleus together. When the muon made its appearance in cosmic rays in 1936, it was at first thought to be Yukawa's meson, but cracks soon started to appear in this cosy picture. The muon interacted very weakly and so could not be the carrier of the strong force. Pions, discovered in the 1940s, turned out to play the role anticipated by Yukawa, but that left the muon in need of an explanation. It was not the only oddity: other particles appearing in cosmic rays were given the name 'strange', so unexpected were they.

In the 1950s and early 60s, QED came into its own as the theory of electromagnetic interactions, thanks in part to CERN's muon g-2 measurement, but it was equally clear that QED did not account for strong interactions, or for the weak interactions exhibited by muons. The 50s had been a decade in which experiment moved faster than theory, leaving the theorists with much to do. It had also been a decade in which the experimental techniques of cosmic ray studies were giving way to the era of the big particle accelerators like the CERN PS and Brookhaven's AGS.

These two machines were in the vanguard of a great transatlantic rivalry in particle physics. This was, and remains, fierce, but it's no ordinary rivalry. Competitors in particle physics chase the same goals, and they all recognise that strong competition helps everyone in the long run: it focuses the mind, and provides that essential ingredient of science – reproducibility of results.

6

IN THEORY

The Standard Model of particle physics is among the most successful and most extensively tested theories in science. It accounts for all of the fundamental particles and the interactions at work between them.

According to the Standard Model, just four basic ingredients are needed to make up all the diversity we see in the matter of the visible universe: they are called up quarks, down quarks, electrons and neutrinos. They all have distinctive quantum properties such as electric charge, weak charge, colour charge and mass, which determine how they interact with other particles. These particles are collectively known as fermions after the Italian physicist Enrico Fermi who first described their behaviour.

Alongside these basic building blocks of matter is another category of particles, bosons, which are named for the Indian physicist Satyendra Nath Bose. Bosons, which include photons of light, behave very differently, mediating interactions between the particles of matter and dictating how matter organises itself, from the very tiny distance scales of atomic nuclei right up to cosmic distance scales. An essential difference between fermions and bosons is that fermions cannot share the same quantum state at the same time and place; they have to have at least one quantum number different from their partners, whereas bosons can overlap in a condensate. It's this distinction that requires particles made of fermions to take up physical space: a necessary prerequisite for matter as we know it to exist.

In today's universe, we perceive four kinds of interactions between particles: gravity, the weak nuclear force, electromagnetism and the strong nuclear force. It is gravity that determines cosmic choreography from the scale of planets orbiting stars to the intricate ballet of galaxies, galactic clusters and the vast entirety of the universe itself. Electromagnetism carries energy to us from the Sun, binds electrons to nuclei to form atoms, and atoms to each other to form everything from paper clips to people and planets. The other two forces, the strong force that holds nuclei together and the weak force that causes them to disintegrate, are confined to the tiny distance scales of the nucleus itself.

To understand how it all works, take the electron. It has electric charge and it has mass, meaning that it interacts via the electromagnetic interaction and gravity. The electromagnetic interaction, which is carried by photons, acts on the electron's charge, while gravity acts on its mass. These two forces are responsible for many familiar phenomena. It's the interaction between our mass and that of the planet that keeps us on the Earth, while electromagnetism brings electricity to our homes through the movement of electrons along wires, and allows us to see things through the interaction of photons, which are particles of light, with our retinas.

Like electrons, up quarks and down quarks also have mass and electric charge, so they feel the electromagnetic and gravitational interactions, but the quarks have colour charge as well, which means that they are also subject to the strong interaction, carried by particles called gluons.

Armed with this kit of parts, we can now start to build atoms. Two up quarks and one down quark glued together by gluons make up a proton. The up quarks each have an electric charge of $+\frac{2}{3}$, while the down quark has an electric charge of $-\frac{1}{3}$, so putting them together gives the proton an electric charge of +1, just the right amount to pair up with an electron, with a charge of −1, to form an electrically neutral atom of hydrogen. If instead of two up quarks, we take two downs and put them together with a single up, we have an object very much like a proton, but with no electric charge: a neutron. Different combinations of protons and neutrons make up the nuclei of all the elements of the periodic table. The strong force holds protons and neutrons together via the exchange of pions, just as predicted by Hideki Yukawa. Electrons orbit nuclei to form neutral atoms, which then agglomerate via the electromagnetic interaction to form complex objects. Most of the nuclei we are familiar with are stable because, although the like charges of the protons repel each other, the strong binding force is much stronger than the electromagnetic repulsion between the protons.

Neutrinos, the particles proposed by Wolfgang Pauli to account for energy conservation in radioactive beta decay when a nucleus ejects an electron, are different yet again. They have only a very tiny mass, even by the standards of the fundamental particles, and the only charge they carry is the weak one. That means that they feel only the weak interaction and gravity, which makes them very slippery indeed.

They can pass through the Earth as if it were not there, and so many of them are produced through the energy-generating mechanisms of the Sun that roughly a billion of them pass through an area the size of your fingertip every second all over the Earth. Because they do not feel the strong or electromagnetic interactions, they do not play a part in the solid tangible structures made up of quarks and electrons, but because they are so copious, they are an important area of study for particle physicists.

The relative strengths of the forces, along with their ranges, are what determine the structure of matter from tiny distance scales right up to the vast expanses of the universe. The strong force's influence is confined to the scale of the nucleus. It is carried by gluons. Photons, on the other hand, have an infinite range: they will just keep going until they interact. That's what allows us to see deep into space. If you were to look up into the southern sky at night, and pick out Proxima Centauri in the constellation of Centaurus, our closest stellar neighbour, your eyes would be capturing photons that had been travelling for about four and a quarter years, and you'd be seeing the star as it was that long ago. Images taken by instruments like the Hubble space telescope reveal a much more distant past, looking back billions of years, almost to the dawn of time. Gravity has an infinite range too, and its hypothetical carriers are dubbed gravitons. Their predicted properties are consistent with the recent observation of gravitational waves. The weak force, which is carried by particles called W and Z bosons, is also short-ranged, with a sphere of influence confined to the scale of the nucleus.

The strengths of the forces vary by a factor of 10^{39} (1 followed by 39 zeroes), with the strong force, as its name suggests, being the strongest, and gravity the weakest.

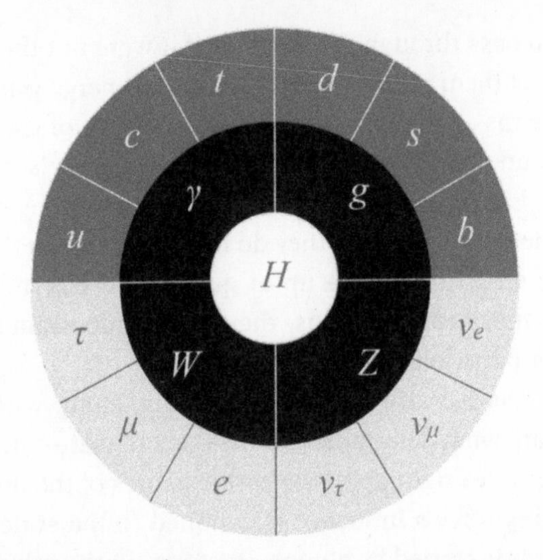

Particles of the Standard Model. The matter particles are in the outer ring. The quarks form the top half of the ring with leptons below. The force carriers form the middle ring, with the Higgs at the centre of it all.

Reproduced by kind permission of Particle Fever, LLC (particlefever.com)

Electromagnetism and the weak force are in between. It is the strong and weak nuclear forces that rule the roost at tiny distance scales, and gravity that determines how the universe behaves at large distance scales. For everything in between, from atoms to people and planets, it is the electromagnetic force that holds things together.

Completing the Standard Model is antimatter. Particles of antimatter are just particles like all the others, but with different properties. For example, an anti-electron, or positron, has the same mass as an electron but the opposite electric charge. Our universe seems to favour the particles we call matter over those that we call antimatter. Why that should be is still unknown.

Today, the Standard Model is firmly established, but as the 1950s turned into the 1960s, its initial foundations were just being laid. The 50s had been a decade in which a range of new experimental discoveries had sent theorists leaping to their blackboards. The 60s would be the decade in which theory started to make sense of it all.

Quarks and aces

The 60s saw an explosion of creativity in theoretical particle physics. By then, CERN's theorists had moved into their new home in building 4 on the CERN campus. Two of the titans of 20th-century theory, Murray Gell-Mann and George Zweig, were independently speculating as to what might lie inside the protons and neutrons that make up atomic nuclei. To Zweig, they were real physical objects that he called aces. To Gell-Mann, they were a mathematical construct that he called quarks. For both, there were three per proton or neutron. Zweig wrote his paper on aces during a one-year fellowship at CERN, but it was not taken seriously by the head of CERN's Theory Division, and so was never published. Nevertheless, although quark is the word we use today, Zweig was right in believing them to be real physical particles.

The road to quarks began in 1961 when Gell-Mann classified particles according to their quantum numbers, rather like Mendeleev had classified the elements according to their properties in the periodic table. By this time, many new particles had been discovered, some with the property of strangeness as had been attributed to some of the particles discovered in cosmic rays, some without; some with electric charge and others without. People were beginning to wonder whether there might be some underlying substructure that would make sense of the complexity, so Gell-Mann plotted them out according to their strangeness and charge.

When the mesons were arranged in such a way, they formed an octet in the shape of a hexagon with two particles at the centre. The same was true for heavier particles, collectively known as baryons, whose spin quantum number was ½, such as protons and neutrons. Baryons with spin 3⁄2 seemed to form a decuplet, but one particle was missing. Just as the periodic table had allowed the existence of unknown elements to be predicted, so this missing place in the baryon decuplet allowed Gell-Mann to predict a particle that he called the omega minus (Ω^-) which was duly discovered at Brookhaven in 1964, leading to a Nobel Prize for Gell-Mann in 1969.

It was also in 1964 that Gell-Mann and Zweig independently proposed their quarks and aces. In analogy to the periodic table, these were the equivalents of the protons and neutrons that account for all the diversity of the elements. The quark model explained all the menagerie of mesons and baryons through different combinations of just three kinds of quarks, though at this point it was not clear whether the quarks were real physical entities, or just a convenient piece of mathematics. Certainly, no one had ever observed a quark.

In the quark model, pions are made up of combinations of up and down quarks and antiquarks, whereas the strangeness of those mesons that had earlier been dubbed strange came down to the presence of a third quark, appropriately named the strange quark. The baryons, which are heavier, are all composed of three quarks.

Despite the success of the quark model, it was some time before experimental evidence showed that the quarks were real particles, and theory explained the reason that nobody had ever seen a solitary quark. The breakthrough came through a series of experiments conducted in the late 60s and early 70s at the Stanford Linear Accelerator Center's End Station A.

The Stanford linac was originally called 'project-M', for monster – and it is just that. A two-mile-long electron accelerator that crosses the San Andreas Fault, it was built in the early 60s and by the end of the decade was delivering 20 GeV electron beams to a switchyard of experimental end stations. In one of the landmark discoveries of particle physics, Jerome Freedman, Henry Kendall and Richard Taylor used one of these beams to look inside the protons and neutrons in liquid hydrogen and deuterium targets, rather as Rutherford had used alpha particles to peer inside atoms. What they found was history repeating itself: the electrons occasionally bounced off at large angles, suggesting that there were small hard objects inside the protons and neutrons, just as Rutherford's alpha particles had revealed the presence of tiny dense nuclei at the centre of atoms. Detailed analysis of the electrons' scattering confirmed the number of these particles to be three, just as the quark model had predicted. A new layer of matter had been unveiled, and a new direction set for the field of particle physics. Friedman, Kendal and Taylor received the call from Stockholm in 1990 to pick up the Nobel Prize for their efforts.

Broken symmetries

In the early 1960s, particle physics had a problem. The long-range interactions, electromagnetism and gravity, could be explained by the theories of the day. General relativity was firmly established as the theory of gravity, while QED had proven its worth in describing electromagnetic phenomena, partly thanks to CERN's muon g-2 experiment; but there was no good theory to account for the short-range weak interaction. The idea that the carriers of the weak force must be heavy, while the carriers of long-range forces were massless could account for the difference, but where would the heavy carriers get their mass? Conceptually the idea of heavy force carriers made sense, but there was no way to reconcile massive and massless force carriers in existing theories, and thereby take the next step towards bringing all the forces of nature together in a single theory.

Newton had taken the first step along that road in the 17th century when he realised that terrestrial and celestial gravity were one and the same thing. Maxwell took the second in the 19th century, bringing electricity and magnetism into the same theory. Now, a century later, the three theorists Sheldon Glashow, Abdus Salam and Steven Weinberg were taking the third step in the long and painstaking journey towards a unifying theory. Working independently, they developed a theory that could unify Maxwell's electromagnetism with the weak interaction on condition that the particles mediating the weak force were massive, limiting their range, while the carriers of the electromagnetic force were massless, allowing them to travel forever, or until they encountered an obstacle. Their theory did not explain the masses of the particles, but it did predict the existence of a neutral carrier of the weak force, called the Z boson, as well as two charged ones, the positive and negative W bosons. The theory they developed is called electroweak, and it earned them the Nobel Prize in 1979.

Before that, however, there was still work to be done: what would account for the difference in masses of the force carriers? The answer relies on the concept of symmetry. Just as symmetry is important in the macroscopic world, so it is in the microscopic world. Macroscopic symmetries are things like reflection in a mirror or rotation about an axis. Take a spinning top of uniform colour, for example. It has mirror symmetry because it looks identical to its reflection, and while it is spinning, it has rotational symmetry because it looks the same however it turns. After a while, however, as the top slows down, wobbles and falls, the symmetry is broken.

At the microscopic level, symmetries, and broken symmetries, play a very important role in all kinds of areas. At the Big Bang, there were just as many particles of matter as there were of antimatter, or in other words, the symmetry between matter and antimatter was perfect. Almost immediately, however, that symmetry was broken, and it's a good thing for us that it was. When a matter particle meets an antimatter particle, both disappear, leaving behind pure energy. In a universe with perfect matter–antimatter symmetry, it would not have taken long for all the matter and antimatter to disappear, leaving nothing behind but a slowly cooling afterglow. The universe would have been a very uninteresting place. But that's not what happened. Instead, a tiny asymmetry appeared, less than one part in a billion in favour of matter. So for every billion particles of matter created at the Big Bang, one survived the primordial annihilation to fill the universe with galaxies full of stars, some of which have planets, on at least one of which intelligent life has developed.

We know this asymmetry is there because we can measure it. By looking out into space, it's possible to measure the amount of matter and the amount of radiation in the form of photons left over from that primordial battle between matter and antimatter, and the photons outnumber the matter particles by ten billion to one. What caused the symmetry to be broken, nobody knows, although the way it's broken is well documented, and requires a diversion along a path that might appear at first sight to be nothing more than an intellectual abstraction.

The microscopic analogues to mirror and rotational symmetry are called parity, denoted P, charge conjugation, C, and time reversal, T. Before the 1950s, these were thought to be exact symmetries for all particle processes, such as the beta decay of cobalt-60, or the disintegration of particles called kaons, which are mesons like pions, but which contain strange quarks. Parity involves looking in the mirror, charge conjugation swapping all particles for antiparticles, and time reversal running a process backwards through time. In a world of perfect symmetry, there would be no difference between cobalt-60 decays and their mirror image. Anti-cobalt-60 decays would be indistinguishable from cobalt-60 decays, and if you ran the process backwards, the laws of physics would remain unchanged.

In 1956, particle physicists were in for a shock when Tsung Dao Lee and Chen Ning Yang proposed experiments to test P symmetry in weak interactions, such as those responsible for beta decay. When experimenters looked closely at the beta decay of cobalt-60, they found that the electrons did not emerge symmetrically, but had a preferred direction. In other words, the mirror image of cobalt-60 decay looked different and the symmetry was broken. Experiments conducted by Jim Cronin and Val Fitch in the 1960s went on to show that the double symmetry of charge conjugation and parity, CP, was also broken in kaon decays, leaving only the triple symmetry of CPT inviolable.

This all sounds very esoteric, but it is one of three conditions listed by the Russian physicist Andrei Sakharov in 1967 as necessary for a matter-dominated universe to exist. Sakharov's first condition states that there must be some process in physics that can change the overall number of quarks or antiquarks, so that from an initially symmetric universe, asymmetry can emerge. The second condition states that the laws of physics must be biased in favour of matter, and that's where CP violation comes in – it favours matter over antimatter, though on its own is not a big enough effect to account for the existence of the universe as we know it. The third Sakharov condition states that this asymmetry must have emerged in a turbulent primordial universe out of thermal equilibrium, since in equilibrium, particles and antiparticles would be produced and destroyed in equal amounts.

Brout, Englert, Higgs, Guralnik, Hagen and Kibble

Understanding the short range of the weak interaction came about through a different broken symmetry first described in 1964 to account for the mass of the particles that carry the weak interaction in the theory put forward by Glashow, Salam and Weinberg. This broken symmetry is every bit as important to us as matter-antimatter asymmetry, because it not only confers mass on the carriers of the weak interaction, limiting their range, but it is also the mechanism that confers mass on all the known fundamental particles, with the possible exception of neutrinos. Without fundamental particle mass, solid objects could not form, and the universe would be just as dull as a universe with perfect matter-antimatter symmetry. Just as there was symmetry between matter and antimatter at the beginning of the universe, so there was symmetry between the force-carrying bosons, and by extension between the forces themselves. As the universe expanded and cooled, this symmetry was spontaneously broken, leading to the forces becoming differentiated and the fundamental particles acquiring their masses. It is this spontaneous symmetry breaking first described by Brout, Englert and Higgs in 1964 that gives rise to the masses of the fundamental particles, but this doesn't mean that all of our mass is made up of fundamental particles: they are all very light, and the bulk of the mass of solid objects ascribed to protons and neutrons comes from the binding energy holding them together.

The first person to bring the notion of spontaneous symmetry breaking to particle physics was Japanese-American theorist Yoichiro Nambu, who took inspiration from a completely different branch of physics: superconductivity. When electrons move along a wire, they encounter resistance, losing energy by jostling with the atoms in the wire, each jostle causing them to emit a massless photon, carrying energy away and heating up the wire. In some materials at very low temperatures, however, electrons travel without any resistance. This is the phenomenon of superconductivity, first observed by Dutch physicist Heike Kamerlingh Onnes in 1911. To cut a very long story short, in 1957 superconductivity was explained by John Bardeen, Leon Cooper and John Schrieffer in a theory now known as BCS theory (from their initials). According to BCS, in superconductors below a certain critical temperature electrons pair up in such a way that the jostling is in a sense cancelled out, and within the lattice of the superconductor photons become massive so that the electrons do not have enough energy to make them. Instead, they sail along the wire unhindered. Nambu wondered whether this notion could be applied to empty space, beyond the confines of a superconducting material, and paved the way for a universe-pervading field that could confer masses upon particles. His work inspired three scientists in Europe to take the next step.

As is so often the way with good ideas, the concept of particle mass generation through symmetry breaking was developed in more than one place at around the same time, two of those places being Brussels and Edinburgh. It was a modest beginning for a scientific revolution, just two short pages published on 31 August 1964 by Belgian physicists Robert Brout and François Englert, and little more than a page from Edinburgh's Peter Higgs on 15 September, but those two papers were set to influence profoundly the development of particle physics right to this day.

All three physicists are careful to give credit to their forerunners, Nambu in particular. Hints of other influences come from the fact that Higgs has been known to call spontaneous symmetry breaking in particle physics the relativistic Anderson mechanism, a reference to the Nobel Prize-winning physicist Philip Anderson who published on the subject in 1963.

Later, in lectures at Imperial College, students of Abdus Salam would be taught about the Kibble-Higgs mechanism, Salam's tribute to his colleague Tom Kibble. Kibble, along with Americans Gerald Guralnik and Carl Hagen, was also working on spontaneous symmetry breaking in the early 1960s, with the three of them publishing work in 1965 that was complete by the time the earlier papers appeared. Kibble went on to publish a paper on his own in 1967, which was well regarded in theoretical physics circles.

History can sometimes make harsh judgements, and when it came time for the Swedish Academy to award the Nobel Prize following the experimental validation of this work in 2012, they chose to award it to Englert and Higgs alone, Brout having passed away in 2011, although the work of all is greatly esteemed in the high-energy physics community. Today, the mechanism these six theorists described is usually referred to as the Brout-Englert-Higgs (BEH) mechanism, while the famous particle bears the name of Peter Higgs alone.

Over the years, all the prestigious prizes in physics have been awarded to different combinations of Brout, Englert, Higgs, Guralnik, Hagen and Kibble. In 1997, the European Physical Society bestowed its High Energy and Particle Physics Prize on Brout, Englert and Higgs. Physicists are fiercely competitive, yet capable of great magnanimity. 'I was delighted to discover that we are sharing the prize,' was Peter Higgs's reaction. 'I get a lot of publicity for this work, but [Brout and Englert] were clearly ahead of me.' In 2004, the Wolf Foundation honoured the same three, while in 2010, all six shared the American Physical Society's Sakurai Prize.

Nambu's insight to apply the concept of spontaneous symmetry breaking to empty space was profound. In the strange quantum world, what we think of as empty space is anything but. Instead, it is a seething soup of particles constantly popping in and out of existence, and it is this structure in the vacuum itself that gives rise to particle masses. The thing that fills the vacuum is what has come to be known as the Higgs field. Some particles interact strongly with this field, others not at all, and it is the strength of the interaction with the Higgs field that determines the fundamental particles' masses. In other words, the carriers of the weak interaction, W and Z particles, are sensitive to the structure of empty space, they acquire a mass and have limited range, whereas the carriers of electromagnetism are not sensitive to the structure of empty space, so they remain massless and have infinite range. This is how the BEH mechanism can accommodate short- and long-range interactions in a single theory. The long-awaited confirmation would eventually come in the form of excitations of the field known as Higgs particles.

The new normal

The postulation and discovery of quarks, coupled with the development of electroweak theory and a plausible mechanism to account for the different ranges of the fundamental particles were huge steps forward in the 1960s, but there was still a problem. The underlying theoretical framework that the theorists of the 60s were using predicted nonsensical results, such as probabilities of more than 100 per cent for given outcomes. It needed to be renormalised, and that would have to wait for a new decade to begin.

In 1971, Gerardus 't Hooft, a student of Martinus Veltman at Utrecht University, published the first of a series of papers by student and supervisor that would rigorously prove the renormalisability of the electroweak theory of Glashow, Salam and Weinberg. They found a way to produce finite, measurable predictions, notably for the masses of the heavy carriers of the weak component of the electroweak interaction, the W and Z bosons.

Renormalisation is a common feature of the quantum field theories used to describe particle interactions.

Take QED, for example, the original quantum field theory developed in the first half of the 20th century to describe electromagnetic interactions. By the 1970s, QED was a well-established part of the physics canon, but it wasn't always that way. At first, the theory appeared to predict nonsensical results for quantities such as the mass of the electron. The underlying reason for this is that in the strange quantum world, energy can be borrowed from the vacuum for a short period of time and then paid back. It is this feature that gives structure to the vacuum, and it also means that an electron is not a simple object, but a seething mass of virtual photons surrounding a bare electron.

By the time the calculations take all the virtual photons into account, they give the electron infinite mass, which is measurably not the case, and it's experimental measurement that holds the key to renormalisation. In a mathematical sleight of hand, by redefining the electron mass in the theory to be the measured value, the infinites in the calculation disappear, leaving a robust theory that precisely describes observations, and makes testable predictions such as the anomalous magnetic moment of the muon, g-2, first measured with precision at CERN in the early 1960s.

Renormalising QED in the 1940s was a huge step forward, earning a Nobel Prize for Sin-Itiro Tomonaga, Julian Schwinger and Richard Feynman in 1965, and so it was for the renormalisation of the electroweak theory. The work by 't Hooft and Veltman put electroweak theory on solid mathematical foundations, allowing the theory to make precise measurable predictions and opening the way to a long and fruitful period of experimental exploration. They too received the Nobel Prize, in 1999. The theorists had risen to the challenge, and the onus was now firmly back on experiment.

NEW KIDS ON THE BLOCK 7

The experimentalists had much to do, but there remained one crucial missing ingredient: a machine that could shake the Higgs particle out of its hiding place in the vacuum. That machine would ultimately prove to be the Large Hadron Collider, or LHC.

As the Standard Model became established, many scientists bet on the discovery of the Higgs particle; but however elegant and enticing the Brout-Englert-Higgs mechanism might be, the only way to be sure that it was right was to find the Higgs. Nature might have chosen to endow the fundamental particles with mass in a different way, so until it was found, the Brout-Englert-Higgs mechanism would remain no more than a compelling idea.

The problem facing the experimental community was that nobody knew quite how much energy would be needed. Despite its great success, the Standard Model did not make a prediction for the mass of the Higgs boson, or indeed for several other quantities. There are, in fact, no fewer than nineteen free parameters in the Standard Model, quantities that have to be measured and plugged in by hand. These are all interlinked, so the better measured each parameter is, the more tightly physicists can pin down the others. Tightening up the Standard Model's parameters in this way shaped the narrative of the search for the Higgs boson for decades to come. As particle accelerators reached higher and higher energies, producing ever more precise measurements, the range of possible Higgs masses became more and more constrained.

By the time the technologies needed for the LHC were ready in the mid-1990s, CERN's Director-General, Chris Llewellyn-Smith, was in the rare and enviable position of being able to guarantee a discovery: the LHC's energy reach would cover the entire range left open for the Higgs. Either it would be found, or whatever other mechanism nature had chosen would make an appearance instead. There was still the question of Fermilab, though. America's flagship energy-frontier laboratory was gearing up for one final push, and although it did not have the energy reach to cover the entire mass range still open for the Higgs, it covered a large part of it.

The long road to the LHC

All that was still far in the future when CERN set up an accelerator research group in 1957. By this time, the SC was already running, and PS construction was well under way, but CERN was thinking much further ahead and it was not long before the accelerator physicists started to dream of colliders. Instead of accelerating a beam of particles and smashing it into a stationary target to produce the particle interactions that physicists wanted to study, they thought that if beams could be circulated in opposite directions and made to collide head on with each other, that would allow much higher-energy collisions to be produced and studied. Another advantage of a collider was that it would not be a one-shot machine. Particle beams could be stored for hours, since only a few particles would collide each time the counter-rotating beams passed through each other. In the LHC, for example, the beams are arranged in bunches, each a few centimetres long and finer than a human hair at the interaction point. Each bunch contains around a hundred billion protons, yet only a few protons collide at each bunch crossing.

The fact that beams would be stored for hours in a collider meant that ultra-high vacuum technologies had to be developed so that beam particles did not get lost to collisions with residual gas atoms inside the beam pipe. Another challenge to overcome was cramming enough particles into each bunch to make the number of collisions meaningful for research.

CERN's accelerator physicists were not the only ones dreaming of colliding beam machines, and the world's first colliders were made elsewhere, at Frascati in Italy, and in the Soviet Union at Novosibirsk amid the frozen wastes of Siberia. The first operational collider was the diminutive Anello di Accumulazione, AdA. Proposed by the Austrian physicist Bruno Touschek, built in Frascati and shipped to France's Linear Accelerator Laboratory in 1962, it produced its first electron-positron collisions in 1963, changing the course of particle physics forever.

The fact that AdA was an electron-positron collider is significant. Charged particles moving through a magnetic field follow curved trajectories. A negatively charged electron will curve one way while a positively charged positron will curve the other, which means that a single set of magnets can be used to hold beams of both on counter-rotating orbits in a circular accelerator. This technique would be used in many, much larger, machines in the future.

It's easy to see the development of particle physics from a western perspective, but by the 1960s, the Soviet Union was also emerging as a powerful force in the field. Just a few months after Touschek and his colleagues began their first experiments with the AdA collider, the VEP-1 electron-electron collider came into operation in Novosibirsk.

CERN was not far behind, and in 1961 the accelerator research group was promoted to the status of a whole division. With two projects on the drawing board, it had grand designs for the future. One plan was for a much higher-energy proton synchrotron – a 300 GeV machine – and the second for a machine called the Intersecting Storage Rings, ISR.

CERN's accelerator physicists knew about the developments in Frascati and at Novosibirsk, and they were also in touch with a group called the US Midwestern University Research Association, MURA, which had come up with the idea of beam stacking: combining bunches to create a sort of superbunch that would contain enough particles to generate plentiful collisions. In CERN's design, the ISR would not be required to accelerate particles, instead, its job would be to take pre-accelerated proton beams from the PS and store them for hours at the same energy inside two intersecting rings, generating collisions at eight intersection points. For it to work, it would need beam stacking, and so work began on CESAR, a now largely forgotten research machine that nevertheless played a crucial role in CERN's history.

CESAR, or the CERN Electron Storage and Accumulation Ring to give it its full title, was designed to test the idea of beam stacking in a small device to pave the way for the ISR, and it began operation in 1963. By 1965, the principle had been proven, and the CERN Council gave the green light to the ISR, which became the world's first hadron collider when it came online in 1971 – hadron being the collective term for composite particles made of quarks and gluons.

Approval of the ISR led not only to a new era of physics for CERN, but also to a new era of international collaboration. The original CERN site was too small to host another big machine, and more land was needed. The natural place to build the ISR was on land contiguous to the existing site. The trouble was that that land was in another country: France. In 1965, the French government made about 40 hectares of French land adjacent to the CERN site available for the construction of the ISR and CERN became an international organisation on the ground as well as by membership. Years later, this led to the unusual situation that a computer programmer working at the western end of the CERN campus could have an office in France and cross the road for lunch in Switzerland.

The ISR was not everyone's favourite machine. By this time, experimentalists had honed their techniques for fixed-target research, and even though the ISR could deliver much higher-energy collisions, equivalent to those available to a 2,000 GeV fixed-target machine, they had put their weight behind the 300 GeV proton synchrotron project. Nevertheless, when the ISR switched on, the experimentalists were there to take data. They found that particles emerged from head-on collisions in all directions, and often at very large angles to the beam pipe. This meant that the detectors they had initially installed, just a few sparse elements around the collision point, were limited in what they could do. Over its operational lifetime, which ended for financial reasons in 1983, the ISR served as a vital training ground not only for accelerator physicists but also for detector builders. It's where they learned to build collider detectors that would completely enclose the collision point, leaving no gaps for emerging particles to escape.

The experience acquired at the ISR by both accelerator builders and experimental physicists would prove to be a good investment for the future. By the time the ISR switched off, the accelerator physicists had developed new techniques, such as placing powerful focusing magnets either side of the collision points to squeeze the beams as small as possible, maximising the number of proton-proton collisions. Those collisions took place inside a detector that completely surrounded the collision point. In short, the ISR had become the model for future colliders.

Keeping cool

Perhaps the most important technique to be tested at the ISR was invented by an unassuming Dutch engineer who had joined CERN in 1956 to work on power converters – devices that take power from the mains and convert it to the form needed to run particle accelerators. In this he was second to none, but Simon van der Meer was also a man who loved puzzles, and he was someone who had the intellect and patience to solve the most complex ones. He would never use two words where one would suffice. But that one word would invariably be the right one, as he demonstrated week after week by completing the *Observer* crossword, despite not being a native English speaker. Puzzle solving was a talent that served him, and particle physics, well.

One of the biggest puzzles in 1970s accelerator physics was how best to marshal loose crowds of particles into dense bunches that could be used for physics: a process known as cooling. This would prove to be vital in machines that collided protons with their antimatter counterparts, antiprotons. Unlike protons, which can be extracted relatively simply from hydrogen, antiprotons have to be made by smashing a proton beam into a target and filtering off the small number of antiprotons that are produced. It's a slow and painstaking process, and the antiprotons are produced with a wide range of momenta. Cooling is the process that tames them, allowing intense beams of antiprotons to be accumulated.

Simon van der Meer first addressed the question of cooling at a meeting in 1968. He devised a technique that essentially involved two components: a detector that would measure the motion of particles at one point on the ring, and a device that would 'kick', or marshal, beam particles into line at another point on the ring. When the beam passed the detector, it would measure the deviation of a sample of particles with respect to the correct orbit and then send a signal across the ring to the kicker device allowing it to nudge the beam. He called the idea stochastic cooling, stochastic meaning random, since on each orbit a random sample of particles is measured to provide the correction. The technique does not correct the deviation of all the particles at once, but over time each particle is nudged towards the chosen orbit and a once unruly crowd becomes a tightly marshalled beam.

It was to be some time, however, before van der Meer's stroke of genius became reality. As with many a puzzle solver, once one puzzle was solved, he would move on to the next one and his colleagues sometimes had a devil of a time convincing him to write up his ideas. But write them up he did, putting pen to paper in 1972. In a footnote, he wrote: 'This work was done in 1968. The idea seemed too far-fetched at the time to justify publication.' Soon after, his doubts were put to rest when stochastic cooling was successfully tested at the ISR. Thanks to the ISR, and to Simon van der Meer, the key challenges of collider physics had been mastered by the end of the 1970s.

Super Proton Synchrotron

If the ISR was the engineers' choice, the 300 GeV proton synchrotron was what the physicists really wanted. They had experience of working in fixed-target mode, and they were worried that the ISR would not produce the collision rate they needed. As a consequence they were prepared to sacrifice the promise of unprecedented energy reach for the comfort of what they knew. Some argued passionately for the two accelerators to be put forward as a single package to the CERN Council, or for the ISR to be dropped altogether. They were fearful that if the ISR were approved, their 300 GeV dreams would die. Nevertheless, Director-General Vikki Weisskopf decided to work sequentially. He would secure approval for the ISR, and then work on getting the 300 GeV project approved. The physicists' fears almost proved to be well founded as the 300 GeV project embarked on a long and convoluted approval process.

The first proposals for the 300 GeV project were put to the Council in 1964, a year before the ISR got the green light. It was big, but it did not involve any new technologies that had not been tried and tested at the PS and the AGS. In some quarters, this was seen as disappointing – if CERN really wanted to fulfil its ambition to be a world-class accelerator laboratory, surely it should be pushing the limits of technology? Those days would come, and although perhaps not evident at the time, both the ISR and the 300 GeV project would have vital roles to play in enabling the huge leaps that CERN would make in physics and accelerator technology in the decades to follow.

There was a general understanding that the 300 GeV machine would not be built in Geneva, which meant that either the CERN Convention would have to be modified, or a new convention for a separate organisation would have to be drafted. In the end, the Convention was modified to allow CERN to operate more than one laboratory, so wherever the machine ended up, the existing CERN Council would govern it. The assumption that the new machine would be built elsewhere also led to no fewer than 22 site proposals, including from the UK, which was knocking at the door of the European Community and brandishing its European credentials at every opportunity.

An additional impetus came from across the Atlantic where a system of national laboratories was being established to replace the earlier structure of labs that were effectively closed clubs of universities working together without giving access to researchers outside the club. MURA had been one of these, and it was a factor in locating the National Accelerator Laboratory in the Midwest outside Chicago in 1967. Under the dynamic leadership of Bob Wilson, the new laboratory had ambitious plans to seize the high-energy frontier with a 500 GeV machine. Renamed the Fermi National Accelerator Laboratory in 1974, Fermilab would soon take over from Brookhaven as CERN's main collaborative competitor in the United States. Collaboration between CERN and the National Accelerator Laboratory started immediately: CERN accelerator scientists were invited to the new lab to help set up the linac.

Rather than spurring on the 300 GeV project, the American development emboldened those who felt that Europe should be pushing the envelope of technology, and calls were made for a radical redesign of the machine. Member state support for building a larger proton synchrotron ebbed away, and the following years were spent rebuilding confidence. A key appointment proved to be that of John Adams as project leader. As architect of the PS, he was one of the most respected accelerator builders around, and he had a mammoth task in front of him. As the American project moved smoothly forward, the Europeans were still in the starting blocks, with the thorny question of a site for the new machine yet to be resolved.

After much closed-door diplomacy, the Swiss government called a meeting in January 1970 to select the site for the new laboratory, and everything seemed to be on track. But then the meeting was postponed at the request of the German government. To many, all seemed lost – but that was to underestimate the ingenuity and determination of John Adams. On 23 January, he wrote to CERN Director-General Bernard Gregory, who had taken office in 1966, proposing to take the bull by the horns and remove site selection from the equation by building the second CERN laboratory adjacent to the first. That, he argued, would reduce costs, since the existing infrastructure could be used to support the new machine. In particular, the PS could be used as its injector. This, he argued, would answer the concerns of those member states that thought the 300 GeV project too expensive, while also ensuring the future of CERN and its Geneva site.

In his proposal, Adams suggested starting with a relatively modest 150 GeV machine, which could be upgraded to 300 GeV later with the addition of extra magnets, or even further if superconducting magnets became available. Adams' plan also took into account the geological constraints, proposing a machine that could be built in a tunnel about 30 metres underground across the Franco-Swiss border adjacent to the existing laboratory.

Over the following months, the project was discussed in increasingly large circles, and was ready for presentation to the CERN Council in 1971. The project was approved, and CERN Lab II was duly established at Prévessin in France with Adams as its Director-General tasked with building a Super Proton Synchrotron, SPS. Across the border, Bernard Gregory handed over the leadership of CERN Lab I to Willibald Jentschke, and CERN found itself in the strange position of having two adjacent laboratories with two Directors-General. Although this situation lasted for only one mandate, the terms Lab I and Lab II endured well into the 1980s. By this time, the small village of Meyrin on whose land the original CERN laboratory was built had expanded into a substantial suburb of Geneva, complete with an Olympic-sized open-air swimming pool a short distance from CERN. To British graduate students at the time, this was affectionately known as Lab III in view of the amount of time they spent there contemplating the intricacies of their theses.

SPS construction went quickly, and by 1976 the new machine was ready. It accelerated beams to 400 GeV on

17 June 1976, exceeding its initial design goal, and achieved 500 GeV by the end of 1978. Meanwhile, Fermilab's machine had achieved 500 GeV on 14 May 1976. That same year, CERN's brief period of organisational schizophrenia came to an end as Lab I and Lab II were reunited under a single management structure, albeit with two Directors-General. John Adams took the title of Executive Director-General, while the Belgian theorist Léon Van Hove became Research Director-General, both with a five-year mandate.

Bridging the East–West divide

The tiny electron collider, VEP-1, had shown that the Soviet Union was rapidly backing up its already impressive capacity in theoretical physics with experimental facilities. In 1956, the Eastern bloc had established the Joint Institute for Nuclear Research, JINR, at Dubna near Moscow as a CERN equivalent for the members of the Warsaw Pact. And while the western Europeans were deliberating about a 300 GeV machine, the Soviets quietly brought the world's highest-energy particle accelerator of the 1960s online in Serpukhov. Starting up in October 1967, the Soviet machine had a beam energy of 76 GeV. Eager to have access to the world's highest-energy beams, CERN signed an agreement with the Soviet Union under which it would provide specialised equipment for the Serpukhov accelerator in exchange for access for European scientists.

The collaboration established in 1967 continues to this day, with JINR and CERN organising joint summer schools for young physicists, and scientists from across the former Soviet Union collaborating in CERN projects. It was at

CERN that early contacts between scientists from East and West Germany took place, and following the collapse of the Berlin wall, many countries from eastern Europe were quick to join CERN, beginning with Poland in 1991. The strong relations that CERN established with both the US and the USSR even allowed CERN to play a small cameo role in the strategic arms limitation talks of the 1980s. Being well connected to both countries, CERN's Director-General at the time, Herwig Schopper – who had taken up the reins in 1981, succeeding Adams and van Hove – was able to offer CERN as a location for preliminary discussions when he learned that the delegations were having difficulty finding neutral ground.

Revolutionising particle detection

In the 1960s, optical devices like bubble chambers were still the detectors of choice for particle physics. Particle tracks would be photographed, and an army of people was needed to sift through and scan the images. In 1964, thoughts were turning to the bubble chambers that would be needed for the 1970s and 80s. As in the 1950s, the choice was between the relatively easy heavy liquid chambers and the more complex ones using liquid hydrogen.

One of the leading proponents of heavy liquid chambers was André Lagarrigue of France's prestigious École Polytechnique. In 1963, at a conference in Sienna, he sketched out a design for a large heavy liquid chamber, and the following year presented a plan for a 17,000-litre chamber to be built as a European project and installed at CERN. It would be used to study the ghostly interactions of elusive neutrinos: then and now a great line of enquiry in particle physics, with the potential to unlock many of the universe's secrets. Lagarrigue and his colleagues argued that such a chamber would be adequate for the physics required of it, and crucially, it could be ready before any more ambitious liquid hydrogen chamber on either side of the Atlantic.

CERN was hesitant. Space for such detectors on PS beam lines was at a premium, and the laboratory wanted to be sure it was making the right choices before committing. As a result, the French agreed to carry the full cost of construction, leaving CERN with just the modest installation costs. It was an offer too good to refuse, and Lagarrigue's heavy liquid chamber was on the map.

In parallel with the French initiative, work was proceeding on a huge liquid hydrogen chamber that came to be known as BEBC, the Big European Bubble Chamber. It began operation in 1973, three years after Lagarrigue's chamber, which started up in 1970 with a healthy head start on the competition.

Lagarrigue's chamber was known as Gargamelle, after a giantess and the mother of Gargantua in Rabelais's 16th-century stories of Gargantua and Pantagruel. Gargamelle ran from 1970 to 1976 in a neutrino beam provided by the PS, and then continued at the SPS before retiring in 1979. Although BEBC would run on into the 80s, Georges Charpak had sounded the death knell for bubble chambers in 1968 with the invention of the multi-wire proportional chamber. Both Gargamelle and BEBC are now on proud display at CERN.

Like Simon van der Meer, Georges Charpak was an old hand at CERN by the end of the 1960s. He'd joined the laboratory in 1959 after a journey that had seen him move from his native Poland to France when he was seven years

At work inside the Gargamelle bubble chamber in August 1970.
CERN

old. During the war, he served in the resistance and spent time in Dachau, before studying at some of France's most prestigious schools and obtaining a doctorate in 1959 for studies on radioactivity. By the time van der Meer was solving the puzzle of herding particles, Charpak was putting the finishing touches to the multi-wire proportional chamber, MWPC, a device that would revolutionise particle physics and many other fields.

Before Charpak's invention, the only electronic detectors in use in particle physics were effectively used as optical devices. Spark chambers are gas-filled volumes containing metallic plates at high voltage. When high-energy charged particles pass through them, they ionise the gas, producing charged electrons and ions, which are attracted by the high voltage plates and produce a visible spark. Charpak's idea was to tame the spark. Instead of just high-voltage plates, he strung an array of anode wires between pairs of plates to generate a uniform electric field across the chamber. Along with a judiciously chosen gas mixture, this replaced the visible spark with an invisible avalanche of electrons picked up by the nearest anode wire. The resulting pulse could be measured electronically, giving a coordinate for the particle that had caused the initial ionisation. As time went on, the same basic technique was refined: anode wires were deployed in a grid to give two coordinates for the passing particles, and by measuring the time it took for the pulse of electrons to reach the end of the wire, the measurement could be further refined.

Over the years this basic technique not only revolutionised the way that particle physicists carry out their research, it has found applications beyond the field, notably in medical imaging.

A glint in John Adams' eye

As the 1970s advanced, with both the ISR and SPS running, CERN was planning its long-term future. With the increasingly long lead times for new facilities, it was necessary to plan ahead to ensure the continuity of the field. No sooner would one new facility be operational than the accelerator physicists and engineers would be thinking of the next. Then, as results started to come in from the existing facilities, answering some questions and posing new ones, the physicists could start to put together the physics case for a new machine. With the advances in theory of the 1960s and early 70s, there was already a strong physics case for a machine that could discover the W and Z particles predicted by electroweak theory. In 1978, the brilliant Italian physicist Carlo Rubbia made a proposal to the CERN Council to build on the lessons learned at the ISR and convert the SPS into a proton-antiproton collider. Such a machine would have the necessary reach, and could be built with a modest investment.

Rubbia argued that CERN had most of the infrastructure to do it already. The PS was already providing protons for the SPS, and PS beams could easily be used to produce antiprotons. All that was needed was a device to store up the antiprotons until there were enough of them to make a beam. Then they could be injected into the SPS in the opposite direction to the protons and made to collide at two points that would be equipped with powerful detectors surrounding the collision point.

Rubbia's plan led to great successes for CERN, and to a whole new line of antimatter research, but while he was proposing the SPS collider, others were looking still further ahead. Nobody really doubted that the W and Z particles would be discovered, but what then? The SPS collider would probably not be able to study them in fine detail, so thoughts turned to a machine that could: an electron-positron collider.

The experiments that had discovered quarks in the 1960s at SLAC had shown that electrons could be used as precision probes of matter. It's a natural extension that electrons and positrons are good for precision collider physics. Being fundamental particles, their collisions are clean and easy to interpret, unlike those of protons and antiprotons, which are composite and produce messy collisions. When interpreting proton-antiproton collisions, the energetic head-on collisions of quarks and gluons have to be disentangled from the rest of the debris emerging from the collision.

The downside of using electrons and positrons, however, is the amount of energy they lose when going round in a circle. All particles resist being forced to change direction, and charged particles shed energy in the form of photons as they do so. This is called synchrotron radiation. The amount of energy carried away by synchrotron radiation is inversely proportional to the fourth power of the particle's mass, and since protons are some 2,000 times more massive than electrons, they lose vastly less energy. This makes proton machines good for reaching the high energies needed for discovery, while electron machines are the machines of choice for precision physics.

Considerations like these led CERN to consider a change of direction. Having focused on proton machines for all of its life, the laboratory chose in 1976 to study the possibility of building a large electron-positron collider that would be able to do high-precision electroweak physics. By the following year, plans were well advanced and consensus in the physics community seemed to be settling on a Large Electron Positron collider, LEP. John Adams, a proton advocate to the core, was not convinced, but he had a glint in his eye. He felt that Europe should be looking even further ahead. 'We should therefore conceive of a complex of machines which should include both electron and proton machines,' he wrote in July 1977, 'and which can be built on the same site and in the same tunnel.' That little glint in Adams' eye would eventually become the Large Hadron Collider.

FROM A NOVEMBER REVOLUTION TO W AND Z 8

The Standard Model of particle physics is really not one theory but two. Electroweak theory describes the electromagnetic and weak interactions, while a theory known as Quantum Chromo Dynamics, QCD, describes the strong interactions. When 't Hooft and Veltman renormalised electroweak theory, QCD was still in its infancy.

Just as QED describes the interactions between electrically charged particles, so QCD describes the interactions between a different kind of charge, called 'colour', hence the 'chromo' of chromodynamics. The idea that the quarks must have an additional feature was already implicit in Murray Gell-Mann's baryon decuplet (see page 70): the omega-minus particle that it predicted contains three apparently identical strange quarks, and since quarks are fermions, that's not allowed. The resolution to this conundrum began in 1970 when Gell-Mann met German theorist Harald Fritzsch at the Aspen Center for Physics in Colorado. They struck up a collaboration that culminated in their 1973 paper with Swiss physicist Heinrich Leutwyler entitled 'Advantages of the color octet gluon picture'. It marked the beginning of QCD and the establishment of the theoretical underpinnings of the Standard Model as the framework describing the universe at microscopic scales.

QCD has a lot in common with QED, but differs in important ways. Like QED, it is a quantum field theory describing boson-mediated interactions between charged fermions: gluons are the bosons and quarks the fermions. There, however, the similarity ends. Unlike electromagnetism, which has two charge states – positive and negative – QCD has three, labelled red, blue and green. Unlike electric charge, which makes its presence felt at macroscopic scales, the colour charge of the strong interaction seems to be confined within composite particles made of quarks bound together by gluons. In other words, all composite particles are colourless, or 'white'. Inside a proton, for example, there will always be one red quark, one green and one blue, and inside a pion, which consists of a quark and an antiquark, there will be a colour and its anticolour. In both cases, the resulting composite particle is colourless.

The reason for colour's confinement comes down to another crucial difference between QED and QCD: the strength of the interaction is weak at extremely short distances and becomes stronger as the particles move apart. This is called asymptotic freedom, and was first described by Frank Wilczek and David Gross, and independently by David Politzer in 1973, work for which they were awarded the Nobel Prize in 2004. The reason for asymptotic freedom is that the carriers of QED, photons, do not have electric charge, whereas in QCD, there is an octet of multi-coloured gluons, which interact strongly with each other. This is what ensures that every composite particle is colourless. If you try to pull a quark out of a proton, the more you pull, the harder it becomes as the interactions between the gluons intensify. It's a bit like pulling on a piece of elastic string – eventually there's enough energy stored that the gluon string breaks, with the stored energy manifesting itself as a quark-antiquark pair. This is how the strong interaction holds protons and neutrons together in atomic nuclei. It's just as Yukawa supposed. As a quark draws away from a

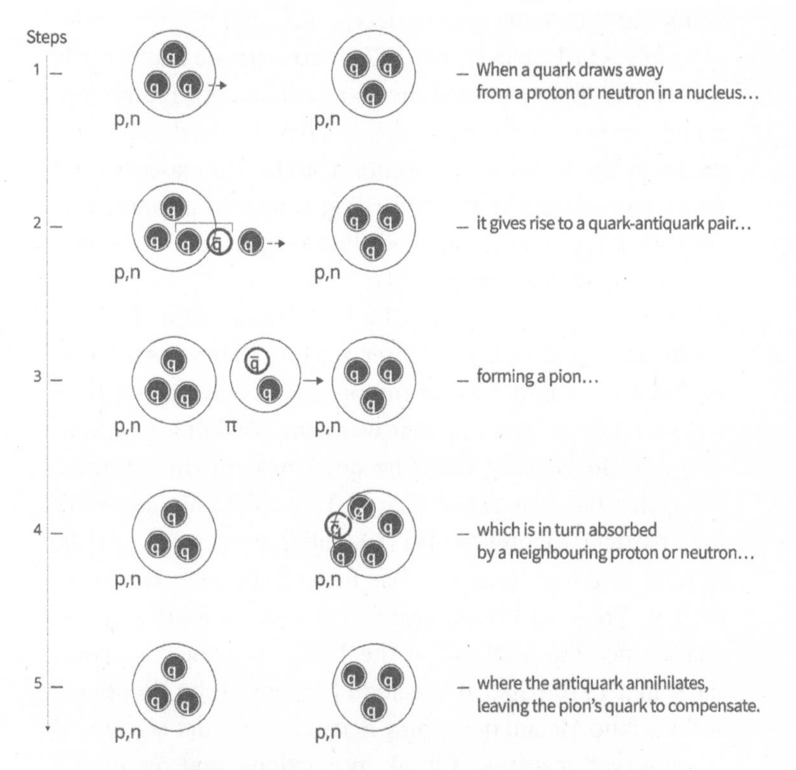

A simplified schematic of how pions mediate the strong interaction between protons and neutrons.

CERN/Daniel Dominguez

proton, for example, it gives rise to a quark-antiquark pair. The new quark remains with the proton, while the antiquark teams up with the escaping quark to form a pion, which in turn is absorbed by a nearby proton or neutron, where the antiquark annihilates leaving the pion's quark to compensate. The net result is that the protons and neutrons are bound together by the exchange of pions, and the interaction is very short ranged.

With theoretical aspects of the Standard Model crystallising and the model gaining acceptance, a drama was about to unfold. On 11 November 1974, two separate teams made a discovery that cemented the Standard Model's reputation. Its impact was so profound that it's remembered in physics circles as the November revolution, and it left experimental physicists around the world ruing a missed opportunity. CERN's ISR, for example, had the energy to make the discovery, but no one was looking.

Two people who *were* looking were Sam Ting at Brookhaven and Burton Richter at SLAC. They both found evidence for a new kind of meson, one much heavier than any seen before, and one that was composed of a new kind of quark. In keeping with the quirkiness of the 'strange' label, this one was called charm. Ting called his discovery the J particle, Richter the Psi (ψ), and in order to give credit to both discovery teams, it has henceforth been known as the J/ψ. There had been speculation about a fourth type of quark since the first half of the 1960s, but the prediction came in a 1970 paper by Sheldon Glashow, John Iliopoulos and Luciano Maiani describing a mechanism that accounted for observed features of weak interactions, and required a fourth quark to exist. The discovery of this new kind of quark added new weight to the quark model, and taken together with renormalisation, put the theoretical developments of the 1960s firmly on the map.

At CERN there was a feeling of gloom at having missed out on the new particle, a feeling that was reinforced three years later when another important discovery slipped through the European laboratory's fingers as a Fermilab experiment announced the discovery of another new heavy meson in 1977 (see page 104). With hindsight, CERN need not have worried about the ISR's legacy. Although in terms of physics, it is remembered more for what it did not discover than for what it did, in accelerator development, it changed the field forever. Without the ISR, it's doubtful that CERN would have taken the bold step to turn the SPS into a collider – a move that would lead to the laboratory's first Nobel Prize in 1984.

Hunting for particles

With the Standard Model firmly established, the onus was now back on the experimentalists to track down its predictions. There was the Higgs boson, of course. Plus the existence of a neutral carrier of the weak force, the Z boson, had been predicted by electroweak theory. Another pair of quarks would soon join the Z on the particle physicists' wish list.

The prediction of a new pair of quarks evolved from the work of one Italian and two Japanese physicists. In the 1960s, Nicola Cabibbo came up with an elegant representation of weak transitions – the way that the weak interaction can transform one quark into another. He represented the transition probabilities in a two by two matrix. It worked well, but it did not account for the CP violation (see page 75) observed in the decay of kaons by Cronin and

Fitch. In 1973 Makoto Kobayashi and Toshihide Maskawa realised that if they generalised the Cabibbo matrix to a three by three matrix, it would account for CP violation. A three by three matrix would, however, require the existence of another, yet heavier, pair of quarks, and by association another electron-like particle and another neutrino. That such additional particles might indeed exist became more clearly apparent in a series of experiments led by Martin Perl at SLAC from 1974 to 1977, which resulted in the discovery of the tau, an electron-like particle some 3,500 times heavier than the electron.

The first of the new quarks, the bottom quark, was found at Fermilab by a team led by Leon Lederman in 1977 in the form of a heavy meson named the upsilon. Fermilab also went on to discover the top quark in 1995. Weighing in at a massive 172 GeV, about the same as an atom of tungsten, the top is the heaviest of the fundamental particles, and its discovery had had to wait until a much higher-energy accelerator than those of the 1970s was available.

For their work, Kobayashi and Maskawa shared the 2008 Nobel Prize in physics with Yoichiro Nambu in a ceremony recognising the hugely important role of broken symmetry in physics. 'Nowadays, the principle of spontaneous symmetry breaking is the key concept in understanding why the world is so complex as it is, in spite of the many symmetry properties in the basic laws that are supposed to govern it,' said Nambu in a speech delivered at the University of Chicago in December 2008. 'The basic laws are very simple, yet this world is not boring; that is, I think, an ideal combination.' With these discoveries, the family of matter particles had grown to twelve, six quarks and six leptons – electrons, muons, taus and their accompanying neutrinos

– but what of the other side of the Standard Model, the force carriers?

The giant awakes

In 1973, while sifting through the thousands of pictures produced by the Gargamelle bubble chamber, physicists came across one picture showing something strikingly unusual. Gargamelle was operating on a PS neutrino beam in CERN's West Experimental area, and while the neutrinos themselves left no traces when they interacted with the liquid target, the charged particles that they dislodged did. What this picture seemed to show was an antineutrino colliding with an electron in the target and continuing unchanged. The electron seems to appear from nowhere, giving rise to a cascade of electrons and positrons moving through the liquid. The only plausible interpretation appeared to be that there was some kind of weak flow, or current, of neutrally charged particles passing between the incident antineutrino and the electron, since no charge was transferred in the interaction. Could it be the Z particle predicted by electroweak theory? There was also another picture showing an antineutrino interaction with a nucleus in the liquid that could also be explained by a neutral weak current carried by a Z boson. Gargamelle scientists made the news public at a series of conference presentations, and on 3 September, their papers were published. For most, this was solid experimental confirmation of the electroweak theory, and would surely have led to the Nobel Prize for Gargamelle's chief protagonist, André Lagarrigue, had he not passed away in January 1975, soon after the result was confirmed by an experiment at Fermilab.

While the discovery of these weak neutral currents at Gargamelle had undoubtedly been the highlight of 1970s physics at CERN, confirming electroweak theory and showing beyond a doubt that the European laboratory was capable of producing world-class results, the 80s would go a step further. All the investment of the 60s and 70s in accelerator technology would pay off, as knowledge gained from the ISR was applied to its former rival for funding, the SPS. The casualty would be the ISR itself, since a decision had to be made: keep the ISR running, or use experience gained from the ISR to convert the SPS into a proton-antiproton collider. There was not enough funding to do both, and, not for the last time, CERN's management was forced to make a difficult choice.

In 1978, the CERN Council gave the nod to the project, and work began on the Antiproton Accumulator, a facility that would enjoy a second coming after the collider project ended, giving CERN a unique low-energy antimatter factory and inspiring everything from writers of fanciful fiction to those wishing to investigate the potential use of antimatter as a treatment for cancer. Stochastic cooling really came into its own in accumulating antiprotons, which are not so easily available as protons. Antiprotons have to be made, and that's a time-consuming process. In CERN's Antiproton Accumulator, stochastic cooling was used to keep them in order as the beam was being built up.

The physics case for converting the SPS into a proton-antiproton collider was also strong since Gargamelle had left unfinished business. Although few seriously doubted electroweak theory on the basis of the Gargamelle result, what Gargamelle had seen remained the proverbial smoking gun: the Z particle itself was still to be found, along with its charged companion the W.

The Tevatron

For a time at the end of the 1970s, it appeared that there might be a transatlantic race for the W and Z bosons, but then Fermilab's Director Robert Wilson took brinkmanship a step too far. Seven experts from CERN, including Carlo Rubbia and Simon van der Meer, were attending a workshop in Berkeley in March 1978 on the subject of producing high-luminosity, high-energy proton-antiproton collisions. Luminosity is a measure of the number of potential collisions per unit area per unit time. Wilson said in his opening address, 'our colleagues at CERN, whom I regard as working for us, are helping to solve the problems and we'll apply the solutions at Fermilab to their dismay.' He was joking, of course: sharing ideas was the lifeblood of particle physics research, and the Europeans duly presented all they knew. If there was to be a race, it would be good for the field, it would help to focus attention, and whoever got the prize, everyone would gain from the new knowledge that it would bring.

Things, however, were not so simple for Wilson. He had difficulty finding the funds to advance Fermilab's proton-antiproton project, and as an act of protest, he tendered his resignation. Much to his surprise, it was accepted, and in 1978 Leon Lederman succeeded him as the lab's director. Lederman immediately reviewed Fermilab's options, and while not abandoning the collider project, he gave priority to getting the fixed-target programme established. A collider capable of TeV-energy collisions, the Tevatron, would follow at a later date. This decision left the field wide open for Europe to discover the W and Z particles.

The CERN Council approved the conversion of the SPS to a collider in 1978, and its two big experiments, UA1, headed by Carlo Rubbia, and UA2 under the leadership of Pierre Darriulat, recorded their first collisions in 1981 at an energy of 900 GeV, or 0.9 TeV. Rubbia's experiment identified its first Z bosons in 1982, and among the first to know was UK Science Minister Margaret Thatcher. She had visited CERN earlier in the year and asked Rubbia to keep her informed. He was as good as his word, writing to Thatcher before the discovery paper was published in 1983. In 1984, Rubbia and van der Meer made the trip to Stockholm. The Nobel committee summed up their contributions in a wonderfully succinct way: 'Simon made it possible. Carlo made it happen.' Experience gained with the ISR, and in particular with stochastic cooling, had given CERN the confidence to make a bold decision, and had set the laboratory on the way to gleaning its first Nobel Prize.

Meanwhile, back in the States, the Tevatron was getting ready to leapfrog the SPS collider to reclaim the high-energy crown. Lederman's decision was a decision for the future, and it set the scene for the Tevatron to make remarkable contributions to particle physics, and to remain at the forefront of the field for a quarter of a century. The Collider Detector Facility, CDF, recorded the world's first 1.6 TeV proton-antiproton collisions on 13 October 1985, rising to 1.8 TeV the following year. In 1991, the SPS, unable to compete, concluded its decade as a collider and reverted to fixed-target physics and being an injector for CERN's new flagship machine, the Large Electron Positron collider, LEP. It was just a few years later, in 1995, that Fermilab was able to announce that the Tevatron's two experiments, CDF and D0, had discovered the top quark, the last of the Standard Model fermions.

A pattern seemed to be establishing itself. After the first fundamental particles to be discovered – the electron in 1897, the muon in 1936, and arguably evidence for the strange quark in cosmic rays – all the fermions had been discovered in the US, while with the discovery of W and Z particles, bosons seemed to be Europe's speciality. As the race to complete the experimental exploration of the Standard Model intensified, would this augur well for the European laboratory? Whatever the future would hold, by 1995, all the Standard Model particles had been discovered, with the exception of one. The Higgs boson, whose discovery would bring understanding of the masses of the fundamental particles, remained as elusive as ever.

THE RACE FOR THE HIGGS 9

As the 1980s advanced, CERN took a new direction. Until now, all of its big particle accelerators had collided protons with fixed targets, with other protons head on, or with anti-protons, but in 1983, civil engineering began for the Large Electron Positron collider, LEP, a giant machine that would be CERN's flagship from 1989 to 2000, and would establish electroweak theory as among the most extensively tested theories of all time. By the time LEP switched off, there was no doubt that electroweak theory was correct, and rested on foundations as solid as Newton's gravity or Copernicus's once-heretical notions of heliocentricity.

LEP's story had begun back in 1977 when opinion in the physics community was starting to focus on thorough validation of electroweak theory. An electron-positron collider with an energy of around 100 GeV per beam, sufficient to produce the W and Z particles in abundance, soon emerged as the machine of choice. Not everyone was convinced. John Adams, for example, wrote a memo to the CERN Directorate on 20 July 1979 expressing his concerns. 'I have been worried for some time now,' he wrote, 'that the overwhelming physics arguments and support for a very large LEP machine are preventing us from considering objectively other strategies for developing our accelerator facilities at CERN.' By this time, the conversion of the SPS to a collider was under way, and LEP, although not yet formally approved, was almost certain to be CERN's next frontier machine. Fermilab, Adams noted, was sticking with protons, having taken the decision to build the Tevatron, and Brookhaven had a plan to build a 200 GeV per beam proton-proton collider called ISABELLE, which was later cancelled. Adams' alternative strategy involved building a smaller, less costly LEP, while simultaneously equipping the SPS with superconducting magnets that would make it competitive with the Tevatron. Adams acknowledged in his memo that he did not expect to gain much support for the notion, and so it proved. But he also advanced the notion of a 10 TeV superconducting fixed-target proton machine being built in the LEP tunnel from the mid-1990s.

Various scenarios were considered for LEP, and hurdles both technical and political overcome. Eventually a design was readied. CERN would build a new linear electron accelerator and an accumulator ring for electrons and positrons, while modifying the PS and SPS to handle electrons and positrons as well as protons and antiprotons, putting these machines well on their way to becoming the world's most versatile particle jugglers. This provided a cost-effective way of using existing infrastructure without compromising other programmes at CERN. Super-cycles of just a few seconds were devised for the PS and SPS, allowing them to serve up protons, antiprotons, electrons and positrons – and later on a range of heavy ions as well – with clockwork regularity wherever at CERN they were needed. The LEP ring itself would be just less than 27km in circumference, buried about 100 metres underground and tilted with a gradient of 1.4 per cent, dictated by the geology of the Geneva basin.

During the last ice age, the Jura mountains that rise to some 1,700 metres above sea level, or 1,300 metres above the city of Geneva, were almost entirely covered in ice, sandwiched between mountain ranges thrown up millions of years ago as the Eurasian and African tectonic plates collided. As a result the geology is challenging: the sedimentary rock, which is stable and largely impermeable, makes a good tunnelling medium, but the porous limestone under the Jura mountains is another story. The position of LEP was duly fine-tuned to situate it at a stable location within the strata of rock rising from Lake Geneva towards the Jura. The result was a tunnel with an average depth of around 100 metres. It would cross the Franco-Swiss border six times, with eight access shafts, four of which would descend to experimental caverns. The shallowest point would be about 50 metres underground, while the deepest, under the Jura, would be 175 metres down.

Among the political hurdles to overcome was ownership of the land. In Switzerland, property owners own their land to a shallow depth, while in France, if you own a house, you own the land it sits on right to the centre of the Earth. That meant different authorisation procedures were required each side of the border, and eventually resulted in everyone who owned property above the ring in France being paid a token amount to cross their land. Later, as green energy started to take off, CERN would face another challenge with the risk that geothermal installations above the ring might accidentally drill into the tunnel. But that would be an issue for another day.

LEP was designed as a two-stage machine, starting with a relatively modest accelerating system based on room-temperature copper accelerating cavities that could deliver a collision energy of 100 GeV and upgrading to a system of superconducting cavities that would boost the collision energy to 200 GeV. In the first phase, LEP I, the energy would be easily sufficient to produce copious numbers of Z particles, whose mass is just over 91 GeV, while LEP II would provide enough energy to produce positive and negative W particles in pairs, as conservation of charge requires. The W particles have a mass of just over 80 GeV. At this time, measurements of the Standard Model's free parameters had not yet allowed the mass of the Higgs to be very tightly constrained, so although the experiments might have a chance of discovering the Higgs particle, precision measurements of the W and Z were the priority.

Council approves LEP

As the June 1981 meeting of the CERN Council approached, CERN's management was quietly confident that their plans for LEP would be approved, but they would have to wait. All twelve of CERN's member states were in favour, but three delegations had not yet received their governments' authorisation to give a positive vote. Since the tradition at CERN was to wait for full consensus, approval was postponed until October at a specially convened meeting.

One of the factors in the decision to build an electron-positron machine at a traditionally proton-oriented laboratory had been to ensure that Europe's flagship facility would be complementary to, though still competitive with, the facilities being advanced in America. The spirit of collaborative competition in particle physics was alive and well. By this time, CERN had a new Director-General, Herwig Schopper, who had begun his five-year term at the start of the year, taking over from John Adams and Léon van Hove.

Civil engineering for LEP got under way in 1983, and was officially commemorated in a ground-breaking ceremony on 13 September attended by Presidents Mitterrand of France and Aubert of Switzerland. At the time, it was the largest civil engineering project in Europe, a mammoth task, and it was accomplished at a time before there were GPS systems to help. The tunnel was mapped out on the surface using a series of geodesic points in the form of yellow pillars dotted around the countryside, all connected by line of sight. The measurements were then transferred underground through a series of boreholes to guide the mammoth tunnelling machines below. When the breakthrough came and the tunnelling teams met up deep under the Jura mountains on 8 February 1988, the massive ring had been completed to an accuracy of just three millimetres.

By this time, the experimental collaborations preparing to do physics at LEP were well established, and because sectors of the tunnel were handed over as soon as they were complete, detector and accelerator installation was well under way. The nature of physics at CERN was rapidly evolving. From the early experiments carried out by a handful of people, collaborations had grown to include dozens of people in the 1980s, and with LEP, they numbered hundreds. The smallest LEP collaboration, OPAL, based on tried and tested technology had around 300 names on its papers, while the largest, DELPHI, employing almost entirely new techniques, had some 700. There were four detectors in all: ALEPH, an acronym contrived from Apparatus for LEP Physics; DELPHI, the Detector for Lepton, Photon and Hadron Identification; OPAL, the Omni-Purpose Apparatus for LEP, and L3, which, under the leadership of Nobel Laureate Sam Ting, was the only LEP experiment to avoid the temptation of choosing a name other than the official designation given by CERN.

LEP's first beams arrived remarkably smoothly. Most big particle physics machines, from the PS to the LHC, take time to work to their design specifications. After all, each is unique, its own prototype, and there's a learning curve to overcome before they deliver their full potential. With LEP, the learning curve was quickly surmounted. On 14 July 1989, as France celebrated the bicentenary of its revolution, a revolution of another kind was happening at CERN as the first 20 GeV beam circulated around the new accelerator. This came as little surprise. Despite its enormous size, LEP I was a relatively simple machine, and the preparations for that first beam had been meticulous: beams had been injected and steered part way round the ring a year before with little trouble, prompting Steve Myers, who was responsible for commissioning the huge machine, to remark ruefully: 'LEP will be more interesting for high-energy physics than for accelerator physics.' Unlike some of CERN's earlier machines, LEP had been the physicists' choice, not that of the accelerator builders.

It took just a month to go from first beam to 90 GeV collisions. It was 13 August 1989, but in the early evening the signs were less than auspicious. One hopeful physicist's phone call to the OPAL control room resulted in the advice: 'Go to bed, nothing's going to happen tonight.' The beams had been lost, and it looked as though nothing much would happen before morning. But then, everything seemed to click. At 21.43, LEP's operators started to accumulate beams. Then at 22.52 the monitor screen flashed up the word 'ramp', in other words, the beams were being accelerated. By 23.00, ramp had given way to 'collide', and at four equidistant points around the ring, eyes were glued to the event display monitors for what seemed to be an eternity. Then, at 23.16, it happened. 'There's one,' said Dave Charlton in the OPAL control room as he spotted the telltale sign of a Z particle decay on the monitor. Jubilant OPAL physicists relayed the word back to the LEP control centre and soon everyone knew that LEP's research programme was under way.

Champagne began to flow and before long, ALEPH and L3 were reporting their first collisions. DELPHI, the most speculative of all the detectors, had still seen nothing. As minutes turned into hours, still nothing was coming up on DELPHI's event display. 'We saw no collisions for about 24 hours,' remembers DELPHI physicist Tiziano Camporesi. 'It was panic mode in the control room, we had to put black and yellow tape on the floor, and you could only cross the line if you were essential to running the detector.' Eventually, there it was: DELPHI's first Z particle. It had been a trivial problem after all: the beams were not quite lining up properly to collide at DELPHI, and the detector had been working fine all along. The LEP era was under way.

It had been a triumphant beginning, but LEP was not alone. In 1980, after taking a sabbatical at CERN, Burton Richter, who'd shared the 1976 Nobel Prize with Sam Ting for the discovery of the J/ψ particle, returned to SLAC and convinced the lab to implement a bright idea that had been kicking around since the previous year. By accelerating both electrons and positrons along the two-mile Stanford linac and bending the beams around arcs to collide, SLAC could have a machine to rival LEP, at least for the study of Z particles – the linac did not have the energy to produce W pairs. The Stanford Linear Collider, SLC, came into operation in April 1989, a significant head start over LEP, but its luminosity was much lower. Nevertheless, by the time LEP was seeing its first collisions, SLAC was showing physics results at conferences based on a sample of 233 Z particle decays.

The first big question for LEP and the SLC was the number of generations of particles that existed. The first generation, from which everything visible in the universe is made, consists of up quarks, down quarks and electrons, along with the neutrino that accompanies electrons in beta-decay, as predicted by Wolfgang Pauli in 1930. Strange events seen in cosmic rays had been explained by a second family consisting of strange quarks, charm quarks, muons and muon-neutrinos, and finally, members of a third family – bottom quarks and tau particles – had also been discovered. Were there more generations to be found? Studying the so-called line shape of the Z particle offered a way to find out, assuming that any as-yet undiscovered neutrino was light enough to be produced by a Z particle decay.

The way the LEP and SLC experiments detected the Z particles produced in electron-positron collisions was by counting the rate of collisions as a function of the beam energy. As the collision energy approached the mass of the Z boson, this rate would grow above a continuum background, signalling the onset of the production of a Z boson. The collisions were detected by observing any of the various particles the Z bosons decay to, for example electrons, muons or quarks. The rate would peak when the collision energy reached the Z mass, and would decrease above that. The graphical representation of the relation between interaction rate and energy is called the Z line shape, and it is sensitive to the number and type of the particles the Z decays to. This includes neutrinos, which escape the detector unseen.

Physicists can model what the line shape would look like for any number of particle generations, and in their models, they assume that the neutrinos remain light in every generation so that the Z particle has enough mass to decay into them. This produced different predictions according to the number of types of neutrino that are invisible to the detector.

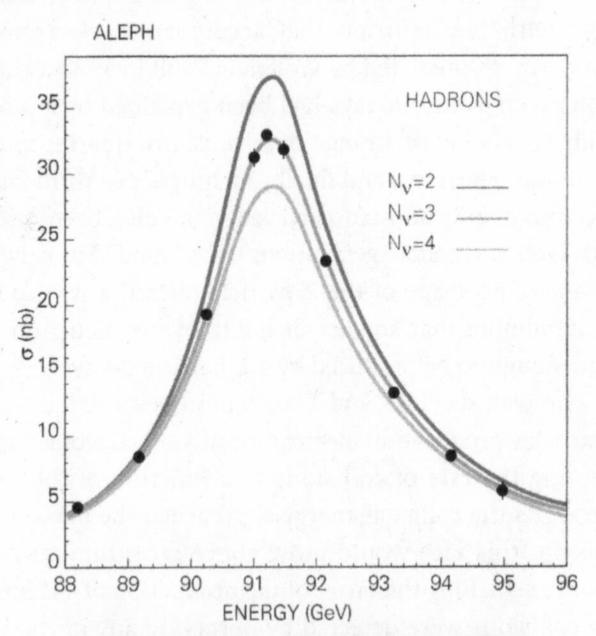

The Z line shape measured by the ALEPH experiment. The data points match the curve for three types of neutrino, indicating that there are three generations of fundamental particles.

CERN

By overlaying the observed line shape on those expected for two, three or four generations, LEP and SLC physicists could see which made the best fit. By August 1989, based on their sample of 233 events, the SLAC physicists were already able to say that the upper limit on the number of particle generations was 4.4.

Meanwhile, those first Z particles detected at LEP had turned into thousands recorded per day, and the LEP experiments were entering the fray. With the first results from LEP scheduled to be made public at CERN in October, SLAC presented an updated result at the European Physical Society's meeting in September in Madrid, saying that the likelihood of there being a fourth family was just 5 per cent based on SLC data. There seemed little doubt, and confirmation would not be long in coming. On 13 October 1989, CERN put out a press release with the unassuming title 'First Physics Results from LEP' containing the news that 'there are no other neutrino types in Nature beyond the three associated with the electron, muon and tau particles'. Why that should be remains an open question.

CERN's press release went on to say that 'until these experimental results from LEP, the number of types of neutrinos had not been determined in a laboratory', just in case anyone was in any doubt. The SLC team may have scooped LEP, but CERN's results were the first that left no doubt, and the European laboratory was determined to have its day in the sun. Over the coming years, LEP would come to dominate where large quantities of data were needed, but the SLC held the advantage on one front: it's easier to polarise beams in a linear collider, and that gives access to physics that no amount of data will bring. The two facilities went on to enjoy a long and friendly rivalry until the SLC switched off in 1998.

Time went on, and before long, everything that LEP could reveal about the Z particle was known, but there was no sign of the elusive Higgs boson. Despite Steve Myers' prognosis, the accelerator physicists had learned many things from the operation of their massive machine. As measurements became more and more precise, the experiments needed to know the beam energy to the level of 20 parts per million, but there was a strange effect that caused the energy measurement to vary by up to 120ppm. Efforts focused on the machine itself until someone at SLAC suggested that the Moon might be to blame. 'We had assumed that something in our hardware was causing these fluctuations – a power supply, or something,' CERN's Lyn Evans told the *New York Times* in 1992. 'But after Dr. Gerhard E. Fischer at the Stanford Linear Accelerator Center in California suggested that lunar tidal effects might be responsible, we conducted experiments that proved beyond doubt that he was right.' The Moon not only causes sea levels to rise and fall, but it also gives rise to Earth tides, causing the size of the LEP ring to vary, and that's what was causing the problem. It later became clear that heavy rainfall, and the level of water in Lake Geneva also had to be taken into account when calibrating this massive, yet extraordinarily sensitive machine.

Another perplexing issue appeared in 1995. There was a periodic perturbance that caused a change in the beam energy of several MeV. But try as they might, no one could find what was causing it, until someone looked at the train timetable and noticed that things seemed to go awry at just the time the TGV train was leaving Geneva on its way to Paris. The railway line passes very close to CERN, and it was realised that some of the current was leaking from the rails, finding its way back to the source through the earth, and in particular through anything conductive like metal pipes or state-of-the-art particle accelerators. Appropriate measures were soon introduced in LEP's calibration.

A new sense of anticipation

By 1995, the four LEP experiments had recorded some 17 million Z particle decays and it was time to upgrade. Superconducting accelerating cavities augmented the copper ones, and LEP's collision energy rose accordingly. There was a sense of anticipation, since this took CERN to the highest energies it had ever seen – perhaps the Higgs would now be within reach. The following year, as the W boson pairs started to flow, the Tevatron reached the end of its first collider run, and Fermilab embarked on an ambitious upgrade for run II. While this would increase the energy slightly, its main focus was on delivering a substantial increase in luminosity, which would increase its sensitivity to rare processes like Higgs boson production. As far as CERN was concerned, however, it left the field wide open until the start of Tevatron run II in 2001.

From 1995 to 1999, the LEP experiments recorded data, diligently looking for signs of the Higgs boson that would complete the Standard Model, or cracks in the theory that would point the way to new physics beyond the Standard Model. By this time, the Standard Model was on very solid foundations, but it was also known to be limited. All those free parameters that had to be measured and plugged in by hand meant that the theory lacked a certain elegance in physicists' eyes – in a perfect theory, everything would come from the equations, leaving nothing to add by hand.

Gravity remained to be incorporated, and there was the question of the universe's dark matter that was known to be there through its influence on ordinary visible matter, but had never been observed directly. This was no mere trifle. At the time, dark matter was thought to account for about three-quarters of the universe, leaving just 25 per cent for all the visible stuff, including us. And things were about to get a lot worse.

As scientists working at the opposite extreme of distance scales to particle physicists observed the farthest reaches of the universe, they made the startling observation that not only was everything moving apart, but the process was somehow accelerating. There must be some kind of energy in the universe driving its expansion. This was rapidly given the name dark energy, and since from Einstein's famous equation, $E=mc^2$, we know that energy and mass are manifestations of the same thing, dark energy could be calculated to account for about 70 per cent of stuff in the universe, relegating dark matter to about 25 per cent and leaving just around five per cent for all the visible stuff. Of that visible stuff, most is gas, leaving just around half a per cent for all the galaxies full of stars and planets, and living organisms such as ourselves. It turns out that ordinary matter is not so ordinary at all. It's actually something very special indeed.

A bump appears

Before LEP retired, there was one final twist in the tale. By the end of 1999, the mass range for the Higgs was being constrained by new measurements, notably that of the top quark's mass, which had been measured at Fermilab, and the most likely mass range for the Higgs had already been explored. That meant that every extra GeV of energy that could be pumped into the collisions could be the GeV needed to shake the Higgs from the vacuum. In 1999, LEP exceeded its design energy, generating collisions at 204 GeV.

In 2000, as LEP started up for its final year of scheduled running, everything was being thrown at pushing the energy as high as it would go. There was nothing to lose; LEP's primary mission was already accomplished, and everyone at CERN was keenly aware that the Tevatron would begin run II the following year. Some of the copper cavities that had been removed were squeezed back into the ring to give a little more accelerating power, pushing the collision energy as far as 209 GeV.

At the end of September, LEP was scheduled to be switched off forever. By this time, civil engineering for the LHC was well under way, and excavations for the huge new experimental caverns were already causing headaches for LEP's operators. The removal of earth from above the LEP ring meant that the machine constantly had to be realigned, and soon LEP operation would be incompatible with progress towards the LHC. Then, on 14 September, CERN put out a press release. 'LEP shutdown postponed by one month,' was the headline.

By this time, the four LEP experiments were adept at combining their data to maximise their chances of observing rare processes, like those that would indicate a Higgs. Because of the rarity of the process, the vagaries of probability meant that while one experiment might see a signal, others could see nothing, but by combining the data, such statistical fluctuations could be ironed out. By the time of the press release, the combined data from the four experiments was showing a bump – which might be an indication that a new particle was being produced – at around 114–115 GeV. Not all the experiments saw it, but it was there in the combined data, and people were taking it seriously. It was a bump rather like that which had signalled the discovery of the Z particle back in the 1980s, but its statistical significance was marginal. Might it be the Higgs? Only more data would tell.

The analysis teams went into overdrive for the next month. There were regular meetings at which all four experiments would present their own results before looking at the combined data. As time went on, OPAL, for example, started to reach the conclusion that there was nothing to be seen by its detector, but crucially, in the combined data, the signal was still there. It was weak, but it had not gone away. Amid a growing clamour to extend the LEP run by a year at the expense of progress towards the LHC, the CERN management was faced with a tough decision. They could run LEP for another year with the risk that the 114–115 GeV signal was no more than a chimera, delaying the LHC in the process, or they could pull the plug, giving the Tevatron a four-year head start on the LHC, whose start date was scheduled for 2005. They took the decision to pull the plug, and at 8am on 2 November 2000, Steve Myers, with great theatricality, did just that, bringing the LEP era to a close on a tantalising cliffhanger.

All eyes were now on the Tevatron, which was about to re-enter the fray. Through the 1990s, both LEP and the Tevatron had made important contributions to knowledge, LEP demonstrating that the number of quarks must be limited to six, as well as putting electroweak theory on firm foundations, and the Tevatron discovering the heaviest of the quarks, the top, in 1995. Although cracks had sometimes seemed to be opening up in the Standard Model, with more data they turned out to be illusory. As the first century of particle physics came to an end, LEP had done the job it was designed to do. There had been no Higgs boson discovery, but the Standard Model was standing strong.

SUPERCOLLIDER! 10

Back in 1982, with Europe on the verge of discovering the W and Z particles and CERN's LEP fully approved, American particle physicists had assembled at Snowmass, Colorado to discuss their next step. Consensus was building that the US should plan a bold leap forward to reclaim the high-energy frontier, and plans were hatched for the Desertron, a machine so big that it would have to be constructed in the vast expanses of desert in the American South-West. Their ambitions were emboldened the following year when the world's first superconducting accelerator, Fermilab's Tevatron, successfully began operation: the Desertron, despite its enormous size, would need superconducting magnets to achieve the collision energy the physicists wanted. Over time, the Desertron became known as the Superconducting Super Collider, SSC. A proton collider, it would have a collision energy of 40 TeV and a circumference of 87 kilometres. The SSC received approval from the Reagan administration in 1987, and a site was chosen at Waxahachie, Texas, just south of Dallas Fort Worth.

American particle physics was on a high. But it was to be short lived.

Meanwhile, Europe's medium-term future was assured with preparations for LEP well under way. It was already clear, however, that whatever followed LEP would not match the SSC for energy. Just as John Adams had wished when discussions about an electron-positron collider began in the 1970s, the LEP tunnel had been built to accommodate a proton collider, and sketches had started to appear depicting a proton machine sitting on top of LEP. Whatever followed LEP would have a circumference of 27 kilometres. That would limit its energy to well below the 40 TeV of the SSC, since even the strongest magnets that people could imagine would not be able to hold such energetic beams on an orbit of that size.

European plans crystallised at a workshop in March 1984, examining options for a proton-proton, or proton-antiproton collider with energy in the range 10–20 TeV. It was there that a plan emerged that would make the much smaller European collider competitive with the American behemoth in some areas. In high-energy physics, it is not just the energy that determines the discovery potential of a machine, but also the luminosity: the more collisions you get, the more sensitive you are to rare phenomena. This is particularly true for proton machines, since the head-on collisions of interest take place between quarks and gluons inside the protons. These particles share the overall energy of the proton: so although quarks and gluons carry just a fraction of the overall energy, sometimes a large part of the proton's energy will be concentrated on a single quark or gluon. In a proton machine, maximising the luminosity brings a double bonus: increased sensitivity to rare phenomena, and boosting the effective energy reach. Of course, if a particle's mass is beyond the energy of a machine, no amount of luminosity will compensate, but a high-luminosity hadron machine in the LEP tunnel could be more competitive than the 14–40 TeV energy difference would seem to suggest. The challenge would be twofold: building an accelerator that could deliver luminosity far higher than had ever been achieved before, and then building detectors that could handle that luminosity. A high-luminosity Large Hadron Collider, or LHC, in the LEP tunnel might well produce multiple collisions at each bunch crossing, but the detector technology of the day would struggle to record much more than one collision per bunch crossing.

High luminosity pointed to a proton-proton collider rather than a proton-antiproton collider, since it is much easier to accumulate large quantities of protons than antiprotons, which have to be made and then stored before being accelerated and collided. On the other hand, a proton-proton collider would need two sets of magnets to bend the counter-rotating beams in opposite directions around the ring. As a result, designs for innovative two-in-one magnets, with two coil windings in the same structure, were pored over.

In spelling out the physics case, the proceedings took on a rather poetic tone: by the time the workshop came to an end, the organisers concluded that 'there is at present a theoretical consensus that the once fashionable desert will actually bloom, but there is no consensus on what flowers exist there.' While it had once been feared that there would be a large range of energy above that already being explored in which there would be no new physics, in other words a physics desert, there was now a plethora of ideas as to what might be found in the 10–20 TeV range. The origin of mass was top of the wish list, but many more questions would also be open to such a machine. What lies beyond the Standard Model, what is the reason for the three families of fermions, and what physics underlies the broken symmetries that are so important in nature? The physics case was compelling, but the main conclusion from the workshop was that there was a lot of work to be done in developing the technologies for high-field magnets, and for detectors capable of handling high luminosity, not to mention the computing resources that would be needed to store and analyse the data. LHC computing is possibly unique in that it assumed that Moore's law, which concerns the regular doubling of computing power over time, would continue to hold for the coming decades.

Accelerator and detector prototyping rapidly got under way and in 1992 at a workshop in Evian, physicists put forward twelve expressions of interest, which rapidly consolidated into four large experimental collaborations. ATLAS, whose name was derived from A Toroidal LHC ApparatuS, and CMS, the Compact Muon Solenoid, would be general-purpose detectors sensitive to any physics that might emerge at the LHC. LHCb would focus on the physics of bottom quarks, and ALICE, A Large Ion Collider Experiment, would concentrate on the collisions of heavy ions in a bid to understand the primordial soup of quarks and gluons, known as Quark-Gluon Plasma, QGP, thought to have existed at the birth of the universe.

In Evian, it looked very much as though a new round of collaborative competition was shaping up across the Atlantic, but then disaster struck. On 21 October 1993, Congress cancelled the SSC in favour of funding the International Space Station. Some $2 billion had been spent, and several kilometres of tunnel dug, but the cost estimates kept rising, and efforts to attract foreign investment had come to naught. The American particle physics community was in a state of shock and global particle physics on a knife-edge: the LHC had not yet been approved by CERN's member states. At the end of 1993, the future of high-energy particle physics was only assured until the research programmes of the Tevatron at Fermilab and LEP at CERN had run their course.

The task of getting the LHC approved fell to Chris Llewellyn-Smith, who took up the position of CERN Director General in 1994, taking over from Carlo Rubbia. Llewellyn-Smith appointed Welshman Lyn Evans as LHC project leader, and it proved to be a winning combination. With a reputation as someone who understood particle accelerators inside out, and was not afraid of taking difficult decisions, Evans had the respect of his team. As the theoretical physicist who had delivered the keynote talk about the potential for LHC physics at the 1984 LHC workshop, Llewellyn-Smith had the trust of the physics community. He was also a consummate diplomat for whom, in the words of his colleagues, doors just seemed to open.

By the end of Llewellyn-Smith's first year in office, he was ready to go to the CERN Council and ask for approval for the LHC. Although the magnets were still in the prototyping stage, they were sufficiently advanced that a cost estimate could be made, and Llewellyn-Smith had a trick up his sleeve in case Council seemed reluctant: he would offer a two-stage approach, starting the machine with only two-thirds of the dipoles, and consequently lower energy. It would cost more in the long run, but would spread the investment over a longer period of time without increasing CERN's annual budget. Council agreed, and Llewellyn-Smith was effusive. 'Today's decision is a major step for the future of High Energy Physics and CERN,' he said. 'Council's decision represents a 20-year commitment to High Energy Physics research...We hope to welcome friends from other countries to participate in the LHC, not only financial participation but even more for their very important intellectual contributions.' This last sentence was crucial: Council had agreed to revisit the decision before the end of 1997, and perhaps to give the LHC the go-ahead as a single-stage project if Llewellyn-Smith could secure sufficient investment from beyond its member states. The signs were already looking good. John O'Fallon, representing the US Department of Energy, congratulated the CERN Council on its decision and invited the Director General and his negotiating team to Washington to work on the details of US participation in the project. A representative of the Japanese Mission in Geneva said that Japan would also be examining the possibility for cooperation with CERN on the LHC.

American renaissance

It had not taken the US particle physics community long to regroup. They still had the Tevatron, and with the demise of the SSC, many had joined LHC collaborations, bringing valuable experience and expertise with them. Llewellyn-Smith and his negotiating team duly accepted O'Fallon's invitation to Washington, and came back with an undertaking that the US would contribute $530 million to the LHC and its experiments. Japan committed 5 billion Yen (roughly $54 million at the exchange rate of the time) in 1995, and there were contributions from Canada, India and Russia.

At its December 1996 meeting, the CERN Council gave the LHC the green light as a single-stage project, albeit within a very tight budget envelope. At the same meeting, Japan pledged a further 3.85 billion Yen ($35 million). The US and Japan went on to collaborate on building the LHC's final focus magnets, Canada provided quadrupoles, India multi-pole magnets, and Russia built the transfer line magnets that would steer beams from the SPS to the LHC. The start date for the LHC was set for 2005, four years after the start of the Tevatron's run II.

For Llewellyn-Smith, it was mission accomplished. His negotiating skills were undoubtedly a factor, but credit is also due to the early pioneers who had drawn up the CERN Convention back in the 1950s. They had produced a stable platform for projects much longer-term than the political cycles of the CERN member states. By the time the LHC was approved in 1996, the so-called CERN model had been successfully deployed at several other European organisations studying fields as diverse as life sciences and astronomy. The idea of creating a CERN equivalent for the Middle East had even been discussed in the CERN cafeteria, leading to an Arab-Israeli meeting in Sinai in 1995. That initiative would eventually become the SESAME laboratory in Jordan, whose members are Cyprus, Egypt, Iran, Israel, Jordan, the Palestinian Authority, Pakistan and Turkey. SESAME was officially inaugurated in May 2017.

The CERN model is a remarkably simple, yet robust model for international collaboration, and it gave many countries the confidence to invest in the LHC: experience had shown that when the CERN Council gives a green light, it's as good as a guarantee that the project will be seen through to a conclusion.

Trials and tribulations

With the LHC approved, global collaborations building particle detectors for it, and the Tevatron gearing up for run II, the future for particle physics at the turn of the century looked bright. After LEP switched off in 2000, CERN's attention turned almost exclusively to the LHC, but there was a rocky road ahead. The problems began in 2001.

By this time, the true cost of the LHC was known. Prototyping was over and industrial production was under way. CERN management reported that the final cost was coming in at about three billion Swiss francs for the accelerator and experimental areas, some 18 per cent above the estimate presented to the Council in 1994. For a project of the scale and complexity of the LHC, this is remarkably good – the history of construction is peppered with projects from the Sydney Opera House to the Scottish Parliament costing many times the initial estimates. But with memories of the SSC's cost overrun, that's not how the CERN Council saw things. When Director-General Luciano Maiani delivered the news, he was greeted with the establishment of an external review committee which tightened CERN's financial procedures. Agreements were put in place to allow CERN to borrow to make good the shortfall so that member states would not have to increase their contributions. Money was also saved by switching off the SPS for a year in 2005. Things appeared to be back on track, with the LHC start-up now scheduled for 2007.

LHC civil engineering caused some challenges, but these were rapidly overcome. At the site of the new cavern for the CMS experiment, the remains of a fourth-century Gallo-Roman villa were discovered in 1998, and the site had to be handed over to archaeologists while its secrets were revealed. 'We had big earth movers, and they had toothbrushes,' says civil engineer John Osborne. 'It was quite interesting to see the difference in excavation.' Coins minted as far afield as London and Rome were found at the site, causing Lyn Evans to joke, 'This proves the United Kingdom, at least during the fourth century, was part of a single European currency.' When work resumed, new techniques had to be developed to build the large caverns the LHC required in such difficult terrain.

The challenges were greatest at CMS, where 60 metres of water-bearing glacial moraine overlaid the stable sedimentary rock, and the rock just above the cavern formed an underground valley where water would drain. 'The CMS shaft leading to the experimental cavern is a bit like the plughole in a bathtub,' says Osborne. While this challenge was being worked out, the CMS collaboration constructed its detector in a vast hall on the surface. The solution the civil engineers came up with was to freeze the moraine as they excavated the shaft descending to the CMS underground areas. To further complicate matters, CMS required two caverns, one for the detector and another for its services, separated by a seven-metre-thick wall. In order to do that, the separating wall had to be built before the caverns were excavated. When all was ready, the detector was lowered gently into the cavern slice by slice using heavy lifting equipment usually reserved for marine salvage operations. The largest slice weighed almost 2,000 tonnes. It made the 100-metre vertical journey at a speed of 10 metres per hour with a clearance of just a few centimetres on each side.

Diametrically across from CMS, close to CERN's Meyrin campus, is the ATLAS cavern, where different challenges lay in wait. ATLAS is much larger than CMS. Measuring 30 by

53 metres with a height of 35 metres, its cavern is the largest ever constructed in the kind of terrain that is found in the Geneva area. To make something so large, the idea of building from the bottom up had to be turned on its head. The roof space was excavated first, and then the concrete roof was constructed and suspended by cables from a gallery above. That provided the stability needed for the rest of the cavern to be excavated beneath it before the walls could be built and the cables de-tensioned so the roof could sit on the walls.

The first technical setback came in March 2007 when one of the final focus magnets built at Fermilab failed a routine test. It turned out that these magnets needed to be more firmly braced inside their cryostats. The solution that the Fermilab engineers came up with was ingenious. They couldn't simply brace the cold mass of the magnet, chilled to –271°C, directly to the room-temperature cryostat that kept the magnet at that temperature because that would introduce a major heat leak. Instead they built long devices that work a bit like shock absorbers and are fitted into the narrow space between the magnet and the cryostat, lengthening the distance between the anchor point on the magnet and that on the cryostat wall. It worked like a dream, and just six months after the incident, all the magnets that needed to be modified were ready.

Angels, demons and black holes

Somewhere in the year 2000 a book landed on Neil Calder's desk. It bore the dedication: 'Neil, Hope you enjoy the novel! Remember, it's fiction!! Best wishes, Dan Brown.' Neil was CERN's press officer, the book was called *Angels and Demons*, and Dan Brown was its little-known author. After a few pages, Neil gave up and passed it on to a junior member of his team to read. The story involved someone stealing a canister of antimatter from CERN to blow up the Vatican. As the author had emphasised in his note, it was very much a work of fiction. In the event, the book was not immediately a great success, but it was a harbinger of things to come.

In 2003, a distinctly more scientifically rigorous book appeared on the market. Entitled *Our Final Century*, it was written by the British Astronomer Royal, Martin Rees, and it examined a number of existential threats to humankind ranging from asteroid impact to a nuclear apocalypse. It also took a look at Pascal's wager, applied to the LHC. Pascal's wager is all about betting on something that you might consider extraordinarily improbable: in its original formulation, the existence of God. Pascal argued that any reasonable person should believe in God since even if you thought that the chances of being right were remote, the consequences of being wrong were terrible.

At the time the book came out, the Internet was buzzing with the notion that the high-energy proton-proton collisions of the LHC might somehow run a tiny risk of creating an Earth-swallowing black hole or worse, and this was the extraordinarily improbable thing that Rees was examining. He concluded that the LHC would be safe, but that didn't stop the online conspiracy theorists. Soon after, the BBC made a drama documentary called *End Day* bearing a remarkable similarity to the scenarios described in Rees's book. The story played out five different versions of one day on which an American scientist wakes up in London and tries to get to the US to switch on a particle accelerator. Each time, he's thwarted by some catastrophe or other, until finally he gets there, runs the gauntlet of protestors and switches on the machine. The last thing we see as the world comes to an end is physicist Frank Close popping up on the screen to say, 'It's great science fiction, but we can sleep easy.'

Meanwhile, Dan Brown had not given up, and when *The Da Vinci Code* became a bestseller, CERN decided that *Angels and Demons,* along with black holes, presented an unprecedented opportunity to put particle physics in the limelight. When *Angels and Demons* started flying off the shelves, the fictitious space plane used in the book to jet about at 11,000mph was front and centre of CERN's website for anyone curious enough to want to know CERN's opinion of the novel. Those clicking discovered the reality of antimatter research at CERN, which, the laboratory pointed out, is far more interesting than the fiction of the book. They could also learn that yes, CERN does make antimatter, but it would take the laboratory 250 million years to produce the quantity stolen in the book. No, CERN doesn't have a space plane. Yes, the LHC's access control system is biometric, but you couldn't fool it by murdering a physicist and using his eye to get in, because only a living eye would work. CERN's web traffic jumped by an order of magnitude overnight.

Angels, demons and black holes helped make CERN a household name around the world, but what is the reality behind these sensational suggestions? To physicists, if the LHC creates an object called a microscopic black hole, it will be a scientific breakthrough next to which the discovery of the Higgs boson would pale. One of the ideas put forward to explain why gravity is so weak compared to the other forces of nature is that perhaps gravity operates in more than three dimensions of space, but the extra dimensions are curled up on themselves at a very tiny distance scale. The consequences of this would be that in our familiar three dimensions, we only see a fraction of the gravitational force, while the rest is trapped in the invisible extra dimensions. If the brute energy of the collisions in the LHC were enough to push particles so close together that they breached the curled-up extra dimensions, then maybe gravity would appear to be as strong as the other forces, and so-called microscopic black holes would be produced. This would be a step on the way to reconciling general relativity and quantum mechanics, the closest thing there is to a holy grail of 21st-century physics. Calculations showed that such objects, which have little in common with cosmological black holes, would rapidly decay, leaving very distinctive signals in the LHC's particle detectors. Most physicists agree that the LHC's chances of producing microscopic black holes are vanishingly small, but many hope that it might.

Theoretical arguments, however, are not needed to decide the particle accelerator variant of Pascal's wager. That can be done on the basis of observation. The Earth is under constant bombardment from space in the form of cosmic rays, mostly made up of protons, many with energies equal to or higher than those of the LHC. When they reach the Earth, they collide with protons or neutrons in the upper atmosphere, effectively making the universe the largest hadron collider there is.

Cosmic rays permeate the universe, and we've been observing them here on Earth for decades. Nothing untoward has ever happened due to cosmic ray collisions in the atmosphere, and no celestial bodies have ever been observed to blink out of existence unexplained. It is these observations that provide assurance that the LHC is safe. Far from doing something unnatural, the LHC simply brings naturally occurring phenomena into the laboratory where they can be studied.

The big fix

In 2008, CERN took the bold decision to do something no laboratory had ever done before. They invited the world's media in to witness the first attempt to circulate a beam. The date was fixed for 10 September 2008. Some 340 news outlets signed up and shared in the excitement of starting up the world's biggest machine. BBC Radio 4 was there for a full day's programming presented from CERN with the rookie presenter Brian Cox as anchor. Every programme was particle physics themed, with the afternoon play being about a particle accelerator opening a rift in spacetime that allowed neutron-devouring aliens through (by this time, CERN had become quite relaxed about the fanciful fiction the laboratory seemed to inspire).

By the time the LHC's day shift began at 7am, the media centre was full to bursting, and journalists turning up on spec were sent to CERN's main auditorium where they could follow CERN's live webcast of the day's proceedings. Following a couple of false starts, the beam was ready for injection by around 9.30. After an hour of carefully threading a beam step by step around the ring, the time came to go for a complete lap. All eyes were on one small display that relayed the signal from the LHC. The display showed a dot when a bunch of particles passed through a beam screen inside the LHC. One dot showed that the bunch had been injected; two would show that it had done one complete orbit. Lyn Evans did a countdown. A dot appeared on the display, and then, after an agonising wait, the second dot appeared. The LHC's first beam had done one complete turn.

A cheer went up in the main auditorium, and as the morning's tension subsided, there was not a dry eye in the house. With the world watching, the LHC had arrived. The TV audience was estimated to be over a billion, and headlines flashed around the world, some proclaiming that 'CERN is the new NASA', able to engage a whole new generation with science. 'Just another day at the office,' as a beaming Lyn Evans remarked as he left for the day.

But the euphoria was to be short-lived. The week following the first beam was marked by thunderstorms producing power outages. Some progress was made in capturing the beam and circulating it for many turns, but then disaster struck. On 20 September CERN issued a statement saying that an incident had occurred at midday on Friday 19 September resulting in a large helium leak into the tunnel, and that further details would be made available as soon as they were known. When those details were known, it was clear that the LHC would be off for a long time. A faulty electrical connection between magnets had caused the superconducting cable to melt and the resulting electrical arc had punched a hole in the pipe carrying the cooling liquid helium, which in turn had evaporated, leading to a build up of pressure that had propagated along the cryostats containing the magnets, damaging several. 'LHC meltdown before first collision' was *Nature*'s take on the incident. 'End of the world postponed,' said the *Daily Mail*.

The work required would not simply be a case of repairing the damage, but also of making the changes necessary to ensure that it could not happen again. Lessons were learned, further interconnects at risk were identified, and work rapidly got under way. 'We have a lot of work to do over the coming months,' said Lyn Evans, 'but we now have the roadmap, the time and the competence necessary to be ready for physics by summer.' As a sign of confidence, the LHC's official inauguration ceremony went ahead as planned in October. At the end of the year, Director Robert Aymar reached the end of his mandate, handing over to Rolf Heuer, whose first task would be to get the LHC physics programme under way.

On 20 November 2009 the beam was back in the LHC, and this time it was there to stay. Three days later, the experiments were showing images of the highest-energy proton-proton collisions ever recorded, and CERN went into its end-of-year break confident that physics could begin in 2010. By this time, the Tevatron was well into run II, and making a final push for the Higgs particle. By the time physics data taking got under way at the LHC on 30 March 2010, the mass range available for the Higgs had been narrowed down to the region of about 114–157 GeV, with a small window around 180 GeV. Much of that was in reach of the Tevatron, and the race was on.

It was not long before the now familiar result was known. The Higgs proved to be just out of reach of the Tevatron, and it was the CERN experiments that claimed the prize as they announced the discovery to an expectant world on 4 July 2012.

Will we all be going to Stockholm?

It was 8 October 2013, Nobel physics prize announcement day, and the smart money was on the Higgs discovery. The big question was, who would get it? Nobel Prizes are shared by a maximum of three people, and there were six involved in the development of the theory, although one had passed away. And what of the LHC and its experiments? The only Nobel that had ever gone to an organisation rather than one or more individuals was the Nobel Peace Prize, so CERN was surely out of the running. There had long been speculation in the scientific community about when the Nobel organisation would change its rules to reflect the changing nature of science. At the time the prizes were established, science was largely a solitary business, but the LHC had involved thousands and it would be impossible to single out just three for the prize. Things were already changing in the 1980s when Carlo Rubbia and Simon van der Meer received their Nobel, but even though their success depended on the efforts of hundreds of people, there was broad agreement with the Nobel organisation's assertion that Simon made it possible while Carlo made it happen.

Speculation at CERN started to mount that 2013 might be the year things changed, as the Nobel organisation tweeted on 4 October: 'The #NobelPrize can be shared between 3 people/organizations. The #NobelPeacePrize has been shared by three in 1994 & 2011.' Were they dropping a hint? Canadian physicist Pauline Gagnon was gleefully preparing what she'd say if CERN won. Until today, she'd say, only two Nobel physics prizes have been given to women, but if CERN got the award, there would be hundreds.

As has become traditional, the 2013 Nobel physics prize announcement was webcast, with the Nobel website saying that the announcement would be made at 11.45 at the earliest. Rolf Heuer was watching it in his office, while screens around CERN were relaying it to crowds of expectant physicists. Eleven forty-five came and went with no news. Then the website said that the announcement would be made at 12.45. Sure enough, an hour later than expected, out came the members of the Swedish Academy. Higgs and Englert it was. Heuer joined the crowds at CERN to congratulate them on their magnificent achievement, and to raise a glass to Peter Higgs and François Englert.

Amidst the euphoria, there was a slight air of disappointment that the Nobel committee had not taken the opportunity to make the change that seems inevitable, since science is increasingly becoming a cooperative affair. One member of the awarding committee revealed to the French news agency, AFP, that there had been a lot of discussion in that extra hour, with at least one person thinking that the prize should have gone to CERN. But it was not to be. CERN and the global particle physics community celebrated the Higgs: the first step on the LHC adventure.

11 WHAT'S THE USE?

Take a candle. Apply the finest brains on the planet to refining it, and you'll get a better candle. What you won't get is an electric light bulb. For that, you need a curious scientist intrigued by a natural phenomenon who investigates it for no other reason than a desire to understand.

In some cases, that scientist might have a notion that their work might one day lead to practical applications. Take Michael Faraday, for example. In 1850, in response to a question about the utility of his work on electricity – an essential prerequisite for the light bulb – he reportedly told William Gladstone, the Chancellor of the Exchequer, 'one day, Sir, you may tax it'. Faraday's confidence was borne out by a remark made by Prime Minister Margaret Thatcher in a 1988 speech to the Royal Society in which she said, 'the value of Faraday's work today must be higher than the capitalisation of all the shares on the Stock Exchange!'

Others, such as Robert Wilson, the founder of Fermilab, chose to defend basic science on its own merits. When asked by Congress what Fermilab contributed to the security of the United States, his answer was, 'It has nothing to do directly with defending our country except to help make it worth defending.' He had a point, but Faraday's message is profound: take any modern-day technology and trace it backwards through time; the chances are that you'll find a curious scientist at its origin. And despite Wilson's lofty statement, Fermilab has, over the years, contributed much of practical benefit to society.

One area in which particle physics has contributed greatly is medicine. Particle physics relies on the acceleration and detection of particles, techniques that are increasingly used in medical diagnosis and therapy. The idea that particles might be useful in medicine goes back right to the origins of the field. When William Röntgen discovered X-rays in 1895, one of the first things he did with them was to produce the world's first radiographic image, of his wife's hand. And when Ernest O. Lawrence invented the cyclotron in the early 1930s, one of the first things his brother John did with it was invent treatments for leukaemia and polycythaemia using radioactive phosphorous made with the machine.

In 1946, Robert Wilson himself took the idea a step further, promoting proton beams as a better treatment for certain cancers for which traditional radiotherapy caused too much collateral damage to healthy tissue. He realised the potential of the technique because, unlike photons or electrons, proton beams deposit most of their energy at the end of their paths in the so-called Bragg peak. This allows deep-seated tumours, or tumours close to sensitive organs, to be targeted with much reduced risk to surrounding healthy tissue.

The first treatments were performed in 1954 when John Lawrence treated patients with proton beams at Berkeley's 184-inch synchrocyclotron. Three years later the same laboratory scored another first by turning helium ions to therapeutic use. Berkeley's pioneering role wasn't confined to the United States. In 1956 Lawrence's friend and colleague Cornelius Tobias was a guest scientist in Sweden where he helped establish a programme of surgery and therapy using protons from the Uppsala University cyclotron.

Proton therapy was pioneered in physics laboratories, but as accelerator, imaging and computing technologies advanced, it gave rise to dedicated hospital-based treatment centres, the first of which opened in 1989 at Clatterbridge in the UK. As proton facilities became established in hospitals, the accelerator research community turned its attention to heavier ions, such as those of helium at Berkeley, since they pack a heftier punch than protons. After Berkeley came Japan's Heavy Ion Medical Accelerator, HIMAC, at Chiba, which was established in 1994, and a facility at the GSI laboratory in Darmstadt, Germany, which treated its first patients in 1997. Just down the road from Darmstadt is a medieval church whose windows celebrate science, with light streaming through stained glass depicting the principles of ion therapy for cancer, as first imagined by Wilson in 1946.

CERN's particular contribution to ion therapy began in the 1990s. By then, particle therapy centres had been established in many more places, notably Harvard and Loma Linda in the USA, TRIUMF in Canada, and the Paul Scherrer Institute in Switzerland. The emerging field had many champions in the particle physics community and CERN embarked on a study to design an accelerator optimised to deliver a steady dose of particles, protons or heavier ions, as medical therapy requires. Baptised the Proton-Ion Medical Machine Study, PIMMS, it has gone on to form the basis of dedicated centres in Italy and Austria. Italy's National Centre for Cancer Hadron Therapy, CNAO, was championed by Ugo Amaldi, son of CERN pioneer Edoardo Amaldi and himself an eminent experimental physicist. It treated its first patients in 2011, while the Austrian facility, MedAustron, came on stream in 2016.

The story of PET at CERN

Antimatter particles, specifically positrons – those positively charged solutions to Paul Dirac's 1928 equation – play a crucial role in medical imaging, and CERN has a long history of collaboration with the medical community in developing their use. The idea that positrons could be a useful tool for medical imaging goes back to the 1950s, but it was not until the mid-1970s that the first positron emission tomography, PET, images started to appear.

The idea sounds like science fiction: by arranging for antimatter to annihilate harmlessly with matter in a patient's body, doctors could trace metabolic function with a precision never before imaginable. The technique works by associating a positron-emitting isotope to a biological tracer that follows a distinctive path in a healthy person, and injecting it into the patient. As the isotope decays, the emitted positrons annihilate with electrons in the patient's body producing a pair of back-to-back photons that escape the body and can be detected, allowing doctors to localise the tracer in the patient.

One of the key developments making this possible was Georges Charpak's 1968 invention of Multi-Wire Proportional Chambers. These allowed charged particles to be tracked electronically, and they went on to revolutionise both particle physics and medical imaging. Charpak chambers have even been deployed in ports to X-ray entire trucks in real time.

Charpak chambers work by detecting the ionisation left behind by charged particles passing through a gas. The photons that need to be detected in a PET scanner, however, have no electric charge, so another development was needed before Charpak chambers could be deployed in medical imaging. It was provided a few years later by another CERN physicist, Alan Jeavons, who built a device that included a plate of lead designed in such a way that the photons would be converted into electrons that could escape from the lead and carry on into the Charpak chamber to be detected. The first PET scan carried out using this device at CERN was in 1977 by a team led by Jeavons, and it was of a mouse. Soon after, CERN physicist David Townsend moved to Geneva's University Hospital, initiating a collaboration between the hospital and the laboratory to build and evaluate a scanner for clinical use.

By the early 1980s, a prototype installed at the hospital had demonstrated the power of the technique in medical diagnosis. Townsend is very clear on the role that CERN played. 'PET was not invented at CERN,' he explained, 'but some essential and early work at CERN contributed significantly.' Over the following years, as new photon detectors were developed for the LEP and LHC experiments, notably scintillating crystals and their readout systems, the interplay between physics and medicine has continued.

CERN in your pocket

The most ubiquitous technology developed at CERN is without a doubt the World Wide Web. First proposed by Tim

Berners-Lee in 1989, the Web built on a pedigree going all the way back to the start of the space race in the 1950s. On 4 October 1957, the Soviet Union put the first artificial satellite, Sputnik 1, into orbit around the Earth. Although the Americans were not far behind, launching Explorer 1 in January the following year, they had been taken off guard and were forced to recognise that the Soviet Union was perhaps not as technologically behind as once thought. As a consequence of Sputnik 1, President Eisenhower established the Advanced Projects Research Agency, ARPA, staffed it with civilians from academia and funded it with the big bucks normally reserved for the military. Later, ARPA would become DARPA, with the D standing for defence, but that was yet to come.

The new agency's mission was to engage in long-range research and development projects that would ensure that the US was not caught napping again, and it had considerable liberty to set its own direction. One of ARPA's first projects was to build a nationwide network linking university computer departments across the country. It was called the ARPANET, and came on stream in the 1960s. By the 70s, it had evolved into the Internet, and had become established as an indispensable tool for researchers not only in the US, but across much of the global research community. By the time Berners-Lee wrote his proposal in 1989, the Internet was firmly established at CERN, scientists were getting used to the idea of having their own personal computers, or at least terminals, on their desks, and there was a thriving community expanding an existing concept called hypertext, whereby you could jump from document to document just by clicking on a link. Berners-Lee's stroke of genius was to put this all together, and that's what he described in his document, entitled 'Information management, A proposal', which he distributed to colleagues and hierarchy in March – to be greeted with a deafening silence.

Berners-Lee's problem was that few could see the point. All the things he described could be done already on the Internet: you just had to know how – and most physicists did. This was also a time when the screens that most scientists were using were simple text-based devices, despite significant growth in graphical user interfaces in the outside world. Berners-Lee could see that the community needed his proposal, but few were as visionary as he was.

Nevertheless, unknown to Berners-Lee until several years later, his supervisor Mike Sendall wrote the words 'Vague, but exciting' on the document before filing it away. One year later, Berners-Lee resubmitted the proposal with nothing altered but the date, and again nothing happened. But in September, things changed when Sendall ordered him a new computer, a NeXT cube, ostensibly to put it through its paces for potential use at CERN, but knowing full well that Berners-Lee would be using it to develop his World Wide Web. By Christmas that year, he'd developed the basic principles of the Web and written the first browser and server. The browser was a sophisticated piece of software, but was confined to the NeXT computers on which it had been written. All of its sophisticated graphical capabilities were invisible to the physicists working on their text-based terminals. Instead, they were initiated to the Web through a line-mode browser written by technical student Nicola Pellow.

In the spirit of the Internet's development right from the early days at ARPA, Web development was a networked global affair, and as a plethora of browsers started to appear, the Web began its inexorable growth. Its place in history was assured in 1993, when CERN made the basic web software available on a royalty-free basis, in line with the CERN Convention's stipulations that the lab should make the results of its work as widely known as possible.

Another technology originally developed at CERN, one that has literally ended up in our pockets, came along when the Super Proton Synchrotron, SPS, was being built in the 1970s. The SPS was the first of CERN's accelerators to have a computerised control system from the start, and two engineers, Frank Beck and Bent Stumpe, set about designing a system that would give it a simple user interface. 'Only a few knobs and switches must control all of the many thousands of digital and analogue parameters of the accelerator,' they wrote in their 1973 proposal, 'and an operator will watch the machine on at most half-a-dozen displays.' They went on to describe a system that would allow the operator to switch between the myriad parameters to be controlled simply by touching a screen.

Beck was aware of the emerging touchscreen technology, but existing devices were clumsy and cumbersome. Stumpe proposed something new: a touchscreen with a fixed number of programmable buttons presented on a display consisting of a set of capacitors etched into a film of copper on a sheet of glass, each capacitor being constructed so that a nearby flat conductor, such as the surface of a finger, would increase the capacitance by a significant amount, allowing the computer to know which button had been touched. This system went into operation in 1976, and continued to be deployed in accelerator control systems for some 30 years, until the ubiquitous mouse took over. In the meantime, they were developed commercially in the field of scientific instrumentation, and became familiar in laboratories around the world.

Beck and Stumpe were quick to see the potential applications for their invention beyond the world of research, showing it at the huge Hannover trade fair in 1977 in the form of the 'drinkomat', a device that played the role of bartender, mixing cocktails. They touted the idea to Swiss watchmakers and in Silicon Valley, but their ideas were ahead of their time. Although the SPS control system was certainly pioneering, touchscreen technology went on to be developed in many places, and the technical connection between Beck and Stumpe's invention and the phones in our pockets or the drinkomat-like technology serving up coffee in CERN's restaurants is not readily apparent. What's without a doubt, however, is that the needs of the fundamental research environment are continually pushing the limits of ingenuity. Particle physics technology is everywhere.

EPILOGUE: WHAT NEXT?

CERN's Large Hadron Collider is a triumph of human ingenuity. It has pushed technology to new and hitherto unimaginable limits. Far from struggling with recording more than one particle collision per bunch crossing – a scenario feared when the LHC was first proposed – the experiments are dealing with dozens. The LHC produces close to a billion collisions in each detector per second, and processes have been perfected to sift out those that might contain interesting physics. The vast computing infrastructure need to store and analyse all this data has kept pace, and the LHC has a research programme mapped out until the mid- to late 2030s, with a major luminosity upgrade scheduled to come on stream around 2025.

Top of the research agenda will be a full understanding of the Higgs particle, with a hope that as we learn more about it, the Higgs might point the way to physics beyond the Standard Model. The study of rare processes, often a good way to unveil new physics, will also grow in importance as the amount of data grows. Already, the LHCb experiment has shown that it can measure processes so rare that they happen only a handful of times in a billion particle decays.

The Higgs was the first of the LHC's major results, but since its announcement, there have been more. The discovery of new exotic mesons, for example, and the discovery of a particle called a pentaquark. The quark model developed in the 1960s predicted that as well as mesons made of a quark and an antiquark, and baryons made of three quarks, there should also be particles made up of five quarks: pentaquarks. Their discovery strengthens the theory of strong interactions further.

When the LHC started up, there were some who hoped that its first significant result would be to provide evidence for a theory known as supersymmetry, SUSY for short, rather than the Higgs particle, but it was not to be. SUSY is a mathematically compelling notion that all the fermions, particles like electrons and quarks, should have a so-called supersymmetric partner that behaves as a boson. Similarly, all the bosons, such as photons or gluons, would have a fermionic superpartner. In this way, the electron would be accompanied by a selectron, the quark by a squark. The photon would be partnered by a photino, and the gluon by a gluino.

In the theory, supersymmetric particles would have been created at the Big Bang along with everything else. Most would have disappeared quite early on, decaying until just the lightest kinds remained. These would be electrically neutral and they would interact weakly with the particles of the Standard Model, making them hard to observe even though they would be abundant in the universe. One of the things that makes SUSY exciting is that the remaining SUSY particles would still be massive, perhaps massive enough to be out of reach of the particle accelerators that came before the LHC, and their mass makes them a strong candidate for dark matter, the invisible stuff pervading the universe and giving it shape.

When the LHC started up, SUSY proponents hoped that the energy reach of the new machine would be sufficient to produce SUSY particles abundantly, taking particle physics beyond the Standard Model, and explaining the nature of dark matter. It would have been quite a discovery: to go beyond the Standard Model and reveal the nature of 25 per cent of the universe at a single stroke. So far, however, physics beyond the Standard Model is proving to be a harder nut to crack. LHC results show that the simplest SUSY models responsible for generating that early optimism are not valid. But the expectation that SUSY remains a relevant property of nature remains, and there is still much room for the LHC to uncover more elusive possible manifestations of it. In parallel, many new theoretical ideas and experimental tests are emerging, going beyond SUSY to extend the Standard Model and reveal the mysterious nature of dark matter.

There will be no shortage of candidate facilities to probe these new theories and, in the longer term, take particle physics beyond the LHC. In the US, Fermilab is transforming itself into a neutrino research laboratory, and in exchange for US contributions to the LHC's high-luminosity upgrade, CERN is building neutrino detectors for Fermilab. Neutrinos provide a potentially fertile hunting ground for new physics, while the high-luminosity LHC upgrade will allow the LHC experiments to make ever more precise measurements, searching for chinks in the Standard Model's armour.

For the longer term, there are plans afoot to build new colliders in Japan and China, while in Europe there are three ongoing accelerator research and development projects coordinated by CERN: the Compact Linear Collider, CLIC; the Future Circular Collider, FCC; and a very long-term project called AWAKE. CLIC uses innovative accelerating technology to pump more energy into beams of electrons than current techniques allow. It could be deployed to build a Higgs factory in a linear tunnel equivalent in length to the diameter of the LHC. The FCC is a much bigger project, more akin to the SSC, but its magnet development programme could also lead to a higher-energy machine in the LHC tunnel. AWAKE is a project to use the field generated in the wake of a proton beam passing through a plasma to accelerate particles. Huge accelerating gradients can be achieved, but there's a lot of work to be done to produce useable beams using such wakefields.

In the increasingly coordinated world of particle physics, European strategy is developed from the grassroots of the field and in consultation with physicists from other regions of the world. Recommendations from the current strategy cycle will be made to the CERN Council in 2020. Whatever the conclusion, one thing is for certain: we still have a great deal to learn about this big wonderful universe we inhabit.

FURTHER READING

CERN and collider technology

The CERN website: home.cern – official home of CERN online.

Collider, Paul Halpern (John Wiley, 2010) – though a little dated now, gives good background on the development of collider technology.

Particle physics

Cracking the Particle Code of the Universe, John W. Moffat (OUP, 2014) – goes beyond the basics of particle physics to explore the complexities beneath.

Introducing Particle Physics, Tom Whyntie & Oliver Pugh (Icon Books, 2013) – graphic guide introduction to the field.

Lost in Math, Sabine Hossenfelder (Basic Books, 2018) – fascinating insider view that the particle physics community has become too driven by mathematics, rather than experiment.

Particle Physics: A Very Short Introduction, Frank Close (OUP, 2004) – a little dated, but packs a lot of background detail into a thin volume.

The Lightness of Being, Frank Wilczek (Basic Books, 2010) – expert exploration of the nature of theoretical physics and the development of modern particle physics.

Who Cares About Particle Physics, Pauline Gagnon (OUP, 2016) – a tour of the Standard Model and what the LHC has contributed.

Quantum physics

Beyond Weird, Philip Ball (Bodley Head, 2018) – concentrates on the widely disputed interpretations of quantum physics in a very approachable fashion.

The Quantum Age, Brian Clegg (Icon, 2015) – a guide to quantum physics with more information than is generally provided on applications, from lasers to electronics.

The Quantum Universe, Brian Cox and Jeff Forshaw (Allen Lane, 2011) – surprisingly in-depth exploration of quantum physics.

Antimatter

Antimatter, Frank Close (OUP, 2009) – a tightly written, quite technical introduction to antimatter.

The Strangest Man, Graham Farmelo (Faber & Faber, 2009) – scientific biography of Paul Dirac, including details of his theoretical work leading to the discovery of antimatter.

The Higgs field and boson

Higgs, Jim Baggott (OUP, 2012) – an excellent in-depth introduction to the Standard Model and the Higgs mechanism.

Massive, Ian Sample (Virgin Books, 2011) – covers both the hunt for the Higgs and the history of colliders.

The Particle at the End of the Universe, Sean Caroll (Dutton, 2012) – a very comprehensive and accessible review of the Higgs particle and the LHC.

INDEX

Page numbers in **bold** refer to pictures

accelerators 25–28, 29
- CESAR 84–85
- circular 26–27, 82–83, 84, 156
- Cosmotron 42
- cyclotrons 26–27, 40, 41, 145
- linear 26, 111, 119–20
- ongoing R&D projects at CERN 155–56
- SC (synchrocyclotron) 46, 51, 53, 56, 60
- SPS (Super Proton Synchrotron) 88–92, 106–8, 111
- strong-focusing 42
- synchrocyclotron 27, 41, 46, 51, 145
- synchrotrons 28, 45, 49–50
- Tevatron 3–4, 6, 92, 107–8, 111, 121, 123, 124, 141
- TRIUMF 27, 146
- Van de Graaf generators 26
- *see also* ISR; LEP; LHC; LHCb; PS; Tevatron; VEP-1; *and specific projects*

aces and quarks 69, 70

AdA (Anello di Accumulazione) 83–84

Adams, John 49, 52, 56, 58, 96–98, 127
- concern about LEP 110–11
- as SPS project leader 90–93

AEC (UN Atomic Energy Commission) 33

AGS (Alternating Gradient Synchrotron) 57, 59–60, 62

ALEPH (Apparatus for LEP Physics) detector 114–15, 116

ALICE (A Large Ion Collider Experiment) 129

Amaldi, Edoardo 38, 49, 147

Amaldi, Ugo (son of Edoardo) 147

Anderson, Carl 17

Anderson, Philip 77

Angels and Demons (Brown, book), and CERN 135–36, 137

antimatter
- particles 17, 68, 87
- research 97

antineutrinos 105
- *see also* Gargamelle (large heavy liquid chamber)

Antiproton Accumulator 106

antiprotons
  creation 87
  production 97
antiquarks 70
ARPA (Advanced Projects Research Agency), later DARPA 149, 150
  ARPANET project 149
asymptotic freedom 100–2
ATLAS (A Toroidal LHC Apparatus) detector 5, 9–10, 129, 134
atomism 12, 13
atoms (*atomos*)
  modern definition 15
  origin 14
  structure 14–16, 66–69
Aubert, Pierre 114
Auger, Pierre 37, 38
AWAKE (Advanced Proton Driven Plasma Wakefield Acceleration), CERN project 156

Bakker, Cornelis 46, 51, 61
Ball, Austin 1
Bardeen, John 76
baryons
  decuplet 70, 99
  structure 70, 154
beam stacking 84, 85
BEBC (Big European Bubble Chamber) 94
Beck, Frank 151, 152
BEH (Brout-Englert-Higgs) mechanism 78–79, 81
Bentvelsen, Stan 9
Berners-Lee, Tim, developer of World Wide Web 149–50
black hole at LHC, speculation of 137–38
Blewett, Hildred 52, 57, 58, 59, 60
Bloch, Félix 47, 49, 50, 51
Bohr, Niels 25, 38
  quantum mechanics, development 23
Bose, Satyendra Nath 64
bosons 64–65
  and fermions 65
  see also W bosons (particles); Z bosons (particles)
bottom quarks 117, 129
  discovery 104
Bragg peak 145
Bragg, William Henry 145
Broglie, Louis de 31–32, 34
broken double symmetry 74–75
broken symmetry (asymmetry) 104, 129
  experiments showing 74–75
  matter-antimatter asymmetry 73, 75
Brookhaven National Laboratory 37
Brout, Robert 2, 3, 8, 11, 75, 76, 77, 78, 81
Brown, Dan
  Angels and Demons (book) 135, 136, 137
  The Da Vinci Code (book) 137
bubble chambers
  heavy liquid 55, 93
  large heavy liquid chamber 93–94
  liquid hydrogen 54–55, 93
  see also BEBC (Big European Bubble Chamber); Gargamelle (large heavy liquid chamber)

Cabibbo matrix 103–4
  CP violation, accounting for 104
Cabibbo, Nicola 103, 104
Calder, Neil 135
Camporesi, Tiziano 116
Cavendish laboratory, Faraday's atom of electricity 15–16
CDF (Collider Detector Facility) at Fermilab 5, 6, 9, 108
CERN (Conseil Européen pour la Recherche Nucléaire)
  announcement of Higgs discovery 7–11, 141
  design and construction 48–51
  east-west collaborations 92–93
  establishment 41–43, 45
  fixed-target physics, conversion to 108
  goals for laboratory, developing 42

internet development 148–51
as model for international collaboration 132
particle accelerators, reasons 41
proton synchrotron, proposal 42, 46
R&D projects, ongoing 155–56
touchscreens, proposal 151–52
see also CESAR; European Organization for Nuclear Research
CERN Convention (document) 43, 44–45, 47
CERN, emergence of 34–43
Auger's initiative 38–39
Council of Representatives of European States, constitution 40–41
intergovernmental meetings 39–41
international laboratory, planning of 40
Kowarski's proposal 35
Rabi's proposal 36–37
see also CERN (Conseil Européen pour la Recherche Nucléaire); European Cultural Conference (Lausanne); European Organization for Nuclear Research
CESAR (CERN Electron Storage and Accumulation Ring) 84–85
Chadwick, James
Chadwick, James, discovery of neutrons 17
charge conjugation 74
and parity, broken double symmetry 74–75
Charlton, Dave 116
charm quarks, J/ψ particles 102
Charpak chambers 147–48
Charpak, Georges 96, 148
MWPC, invention of 94–95, 147
classification of particles **68**, 69–70
CLIC (Compact Linear Collider) CERN project 156
Close, Frank 137
CMS (Compact Muon Solenoid) detector 5, 6, 10, 129
construction 133–35
CNAO, National Centre for Cancer Hadron Therapy 146–47
Cockcroft, John 46
colliders
electron-electron 84
electron-positron 83, 97, 98
hadron at ISR 85
interactions 26–27, 82–84
long-term planning 96–98
proton-antiproton 97–98, 107, 127
proton-proton 111, 127
proton vs. electron, uses 98
see also accelerators
colour charge (gluons) 64, 66, 99
in composite particles 100–1
confinement 100–1
cooling
process 86–88
stochastic 87–88, 106, 108
technique (van der Meer) 87–88
Cooper, Leon 76
Council of Representatives 40–41
Courant, Ernest 42
Cowan, Clyde 51
Cox, Brian 139
CP (charge conjugation and parity) 74–75
violation 75, 103–4
CPT (charge conjugation, parity, time reversal) 75
Cronin, Jim 74, 103

D0 (Fermilab) 5, 6, 9, 108
Dahl, Odd 38, 42, 45, 46, 49
Proton Synchrotron, proposal 42
Dakin, Samuel ffrench 50
dark energy 122
dark matter 122, 155
DARPA (Defence Advanced Projects Research Agency), earlier ARPA 149
Darriulat, Pierre, UA2 experiment 108
Dautry, Raoul 31, 32, 33, 34

Da Vinci Code, The (Brown, book) 137
DELPHI (Detector for Lepton, Photon and Hadron Identification) 114–15
  Z particle detection delay 116
Democritus (atomism, atomic theory of the universe) 12, 13, 23
Desertron 126
  see also SSC (Superconducting Super Collider)
detectors *see* ALEPH; bubble chambers; CDF; cooling; DELPHI; L3 detector; OPAL
Dirac, Paul 17, 147
down quarks 64
  colour charge (gluons) 66

Eddington, Arthur 23
Einstein, Albert 22, 23, 24, 25, 29, 31
  papers by 21
Eisenhower, Dwight D 149
electromagnetic interactions 60–63, 65, 66, 68, 80, 99
electronvolt (eV)
  definition 27–28
  mass in 28
electroweak theory 72
  experimental confirmation 105–6, 110
  renormalisation 79–80
  see also Gargamelle (large heavy liquid chamber)
End Day (BBC, drama documentary) 136–37
energy measurements, causes for fluctuations at LEP 120–21
Englert, François 2, 3, 8, 10, 11, 75, 76, 77, 143
  pictured with Peter Higgs **12**
European Council for Nuclear Research 41, 45
  see also CERN (Conseil Européen pour la Recherche Nucléaire)
European Cultural Conference (Lausanne) 30, 32, 34, 38
  see also CERN, emergence of
European Organization for Nuclear Research (Organisation Européenne pour la Recherche Nucléaire, OERN) 19
  CERN, referred to as 19, 44, 50
  governing structure, development of 46–47
  ratification, process of 44–46, 50
European post-war cooperation on physics 30–32, 33
Evans, Lyn 3, 120, 134, 139, 140, 141
  LHC project leader 130
experiments
  ALICE 129
  ATLAS 5, 9–10, 129, 134
  CDF (at Fermilab) 5, 6, 9, 108
  CMS 5, 6, 10, 129, 133–35
  D0 (Fermilab) 5, 6, 9, 108
  LHCb 129, 153–54
  OPAL 1, 114, 115–16, 124
  see also specific experiments

Faraday, Michael 15, 144, 145
FCC (Future Circular Collider), CERN project 156
Fermi, Enrico 18, 64
Fermilab 3, 5
  fixed-target program 107
  as neutrino research laboratory 155
  proton-antiproton project 107
fermions 64
  and bosons 65
Ferranti Mercury (computer) 55
Feynman, Richard 80
Fidecaro, Giuseppe
  pion decay directly into electron, discovery 53
Fischer, Gerhard E 120
Fitch, Val 74, 104
five sigma, significance of 8–9
forces, fundamental
  gravity, discovery 19–20
  unification theory of 20–21, 29
  see also strong interactions/forces; weak interactions/forces

Friedman, Jerome 71
Fritzsch, Harald 99
fundamental particle mass, importance of 76

g-2 muon
 experiment 72
 magnetic moment 80
 measurement 62
 property 60–61
Gagnon, Pauline 142
Galilei, Galileo 110
Gargamelle (large heavy liquid chamber) 93–94
 confirmation of electroweak theory 105–6
 PS neutrino beam 105
 study of neutrinos 93–94
 at work inside **95**
Geiger, Hans
 atomic nucleus, discovery 16
Gell-Mann, Murray
 classification of particles 69–70
 decuplet 70, 99
 omega minus, prediction 70, 99
 quarks 69
general relativity 29
 for GPS systems 23
 and quantum mechanics, reconciliation 138
 test of 22–23
Gianotti, Fabiola 10
Gladstone, William 144
Glashow, Sheldon 75, 79, 102
 electroweak theory 72
gluons (colour charge) 64, 66, 99
 see also colour charge (gluons)
Goward, Frank 38
gravity 65
Gregory, Bernard 90, 91
Gross, David 100
Guralnik, Gerry 8, 10, 11, 75, 77, 78

Hagen, Carl 8, 10, 11, 75, 77, 78
Hahn, Otto 18
Halban, Hans von 33
heavy force carriers 72
Heisenberg, Werner 24, 40, 41
Heisenberg's uncertainty principle 24
Heuer, Rolf 8, 10, 141, 142, 143
Higgs boson 79, 109
 announcement of discovery 7–11, 141
 'bump' at LEP experiments 122–24
 looking for at LEP 121
 mass 4, 5
 search for 81–82
Higgs field 78
 particle sensitivity to 79
Higgs, Peter 2, 8, 9, 10, 11, 77, 78, 143
 pictured with François Englert **12**
HIMAC (Heavy Ion Medical Accelerator) 146
Hooft, Gerardus 't 79, 80, 99
Hove, Léon Van 92, 93, 114

ICHEP (International Conference of High Energy Physics) 6, 7, 8
Iliopoulos, John 102
Incandela, Joe 10
internet, evolvement 148–52
ISABELLE (proton-proton collider) 111
ISOLDE, online isotope separator 61
ISR (Intersecting Storage Rings) 84, 96–97, 108
 approval 88
 benefits 85–86, 89

Jeavons, Alan 148
Jentschke, Willibald 91
JINR (Joint Institute for Nuclear Research, USSR) 92
 CERN collaboration with 92
Joliot, Frédéric 33
J/ψ particles 102, 116

kaons 74–75
 CP violation 103
Kendal, Henry 71
Kibble-Higgs mechanism 77
Kibble, Tom 8, 75, 77, 78

Kluger, Jeffrey 11
Kobayashi, Makoto 104
Kowarski, Lew 33, 38, 50
  acquisition of heavy water 32
  laboratory for fundamental research, proposal 35
  Operation Swallow: The Battle for Heavy Water (film) 32–33

L3 detector 115, 116
Lagarrigue, André 105
  proponent of Gargamelle 93–94
Lawrence, Ernest O 26, 145
Lawrence, John (brother of Ernest) 145–46
Lederman, Leon 104, 107, 108
Lee, Tsung Dao 74
LEP (Large Electron Positron Collider) 98, 108
  approval 113, 126
  'bump' in experiments 123–24
  closure 124
  collaborations 114
  concept 110
  considerations for building 111–12, 113
  construction 114
  first operation 115–16
  LEP I 113, 115
  LEP II 113
  and SLC, advantages compared 119–20
  its story 110–12
  upgrading of 121
  Z particle decay, recorded 119, 121
  Z particles, detection 116–17
Leucippus (concept of atomism) 12
Leutwyler, Heinrich 99
LHCb
  focus 129
  measurement of rare processes 153–54
LHC (Large Hadron Collider) 1, 9, 81, 83
  challenges during construction 133–35
  conception and approval 128–32
  early experiments 129
  first beam circulation 139–40
  helium leak, consequences 140–41
  history 2–3, 4–6
  international collaboration 131–32
  microscopic black hole, speculation 137–38
  safety of 138
LHC (Large Hadron Collider), future
  Higgs particle, full understanding 153
  luminosity upgrade 153, 155
  rare processes, study of 153–54
Livingston, M Stanley 42
Llewellyn-Smith, Chris 82, 132
  approval of LHC, responsibility for 130–31
Lloyd-Wright, Frank 48, 60
luminosity
  challenges of high 128
  definition 107
  importance 127–28
  LHC upgrade schedule 153
  proton-proton vs proton-antiproton collider 128
  use in LHC 128, 155

McMillan, Edwin 28
Maiani, Luciano 102, 133
Manhattan Project 34
Marić, Mileva 21
Marsden, Ernest, discovery of atomic nucleus 16
Maskawa, Toshihide 104
mass generation through symmetry breaking 77
matter-antimatter 74, 75
  asymmetry 75
  symmetry 73, 76
  and tracing metabolic function 147
matter-dominated universe, conditions for 75
Maxwell, James Clerk 20, 29, 72
medical imaging, PET 147, 148
medical particle treatments 145–48

Meer, Simon van der 86, 88, 94, 95, 107, 108, 142
  cooling technique 87
Meitner, Lise 18
Mendeleev, Dmitri 14, 20, 69
mesons (mesotrons) 154
  J/ψ particles 102
  kaons 74
  production 46
  proposal 62
  structure 70
  upsilon 104
Mitterrand, François 114
Monte Carlo techniques, applications 55
muon-neutrino interaction 117
muons 53
  g-2 property 60–61
  and pion decay process 53
  and pions 62
  see also g-2 muons; QED (quantum electrodynamics)
MURA (US Midwestern University Research Association) 84, 89–90
MWPC (multi-wire proportional chamber) 94–96, 147
Myers, Steve 115, 120, 124

Nambu, Yoichiro 76, 77, 78, 104
NATO (North Atlantic Treaty Organization) 31
Nescafé tin 58–59
neutrinos 64, 66–67
  PS neutrino beam 105
  research and Fermilab 155
  study of 93–94
  and Z line shape 118–19
neutrons, structure of 66
Newton, Isaac 12, 19, 20, 72, 110
NMR (nuclear magnetic resonance) 36
Nobel Prize 2013, expectations at CERN 141–43
November revolution 102

OERN *see* European Organization for Nuclear Research
O'Fallon, John 131
omega-minus particle 70
  conundrum resolution 99–100
Onnes, Heike Kamerlingh 76
OPAL (Omni-Purpose Apparatus for LEP) detector 1, 114, 115–16, 124
  Z particle decay, detection 116
Operation Swallow: The Battle for Heavy Water (film) 32
Oppenheimer, Robert J 33
Our Final Century (Rees, book), Pascal's wager 136, 137, 138

parity 74
  and charge conjugation, broken double symmetry 74–75
particle classification 69–70
  Standard Model **68**
particle interactions/forces 65–67
  ranges of interactions 67
  strength of forces 67
particle mass, generation through 75–76, 77
  see also spontaneous symmetry breaking
particle physics
  aim of 13–14, 15
  benefits of 145–52
  long-term research 155–56
  uses in medicine 145–48
  and World Wide Web 148–52
particles
  classification 69–70
  discovery 15–18, 62
  hunt for 103–5
  J/ψ 102
  number of generations, identification and study at LEP and SLC 117–19
  strange 62, 70
  see also quarks
particle therapy centers
  CNAO, cancer hadron therapy 146–47
  MedAustron, ion beam therapy 147
  Paul Scherrer Institute 146

Pascal, Blaise 136
Pascal's wager, applied to LHC 136, 138
Pauli, Wolfgang 18, 51, 66, 117
Pellow, Nicola 150
pentaquarks, discovery of 154
periodic table
  complexity 17
  formulation 14, 20
Perl, Martin 104
Petitpierre, Max 51
PET (positron emission tomography) scanners 147, 148
Philosophiae Naturalis Principia Mathematica (Newton) 20
photons 62, 65
  light 64
PIMMS (Proton-Ion Medical Machine Study) 146
pions
  decay measurements at SC 55
  decay process for negative 53
  formation 102
  function 66
  and muons 62
  structure 70, 100
Podolsky, Boris 24
Politzer, David 100
positrons (anti-electrons) 17, 68
proton beams, medical treatments with 145
protons, structure 66
PS (proton synchrotron) 42, 46, 62, 85
  accelerating RF cavities 56
  beam capture 54, 56
  beam management 56–57, 94
  beam transition 56–59
  as injector 91, 97
  modification 111
  neutrino beam 105
  particle detection system 54
  start-up 52, 56–59
  team pictured in control room **59**
  see also transition point (at PS)

QCD (Quantum Chromo Dynamics) 99
  asymptotic freedom, reasons 100–2
  charge states 100
  and QED, differences 99, 100–2
  see also omega-minus particle
QED (quantum electrodynamics) 62–63
  muon g-2 property, prediction 60–61
  and QCD, differences 99, 100–2
  renormalization 80
  and strong/weak interactions 63
  see also strong interactions/forces; weak interactions/forces
QGP (Quark-Gluon Plasma) 129
quantum mechanics 29
  consequences 23–24
  development 23
  and need for data 25
quarks
  anti 70
  bottom, discovery 104
  charm 102, 117
  discovery and confirmation (SLAC) 71, 97
  down 64, 66
  experimental evidence 71
  model of 70–71
  penta, discovery 154
  strange 70, 74, 99, 102, 109, 117
  top 104, 108, 122
  up 66
  see also specific quarks

Rabi, Isidor 19, 36–37, 40
Rees, Martin (*Our Final Century*, book) 136, 137
Reines, Frederick 51
renormalisation 79–80, 99, 102–3
research at LHC, future plans 153–54
Richter, Burton 102, 116
  ψ particle 102
  see also Ting, Sam
Röntgen, William 145
Rose, François de 33, 35, 40
Rosen, Nathan 24

Rougemont, Denis de, on European post-war cooperation and science 30–31
Rubbia, Carlo 130, 142
proton-antiproton collider, proposal 97
UA1 experiment (identification of Z bosons) 107–8
see also SPS (Super Proton Synchrotron)
Rutherford, Ernest 71
alpha particles, discovery 16
cyclotron, invention 145
D-shaped hollow electrodes 26
protons, discovery 17

Sakharov, Andrei 75
Salam, Abdus 72, 75, 77, 79
Scherrer, Paul, particle therapy center 146
Schnell, Wolfgang
Nescafé tin 58–59
see also PS (proton synchrotron)
Schopper, Herwig 93, 114
Schrieffer, John 76
Schrödinger, Erwin 24
Schrödinger's cat 24
Schwinger, Julian 80
SC (synchrocyclotron) 27, 41, 46, 51, 56
muon g-2 property, measurement 60–61
start-up 53
as supply to ISOLDE 61
Sendall, Mike 150
SESAME laboratory (Jordan) 132
Sinatra, Frank ('Three Coins in the Fountain') 44
Skinner, Herbert 39
SLAC (Stanford Linear Accelerator Center)
discovery of quarks 71, 97
particle generations, upper limit 119
Z particle decays, results 117
SLC (Stanford Linear Collider) 117
and LEP, advantages compared 119
Smith, Lloyd 60
Snyder, Hartland 42
special relativity 21–22
spontaneous symmetry breaking 76, 77–78, 104
BEH (Brent-Englert-Higgs) 78, 79
SPS (Super Proton Synchrotron) 103, 151
CERN Lab II, establishment 91
conversion into proton-antiproton collider 97, 106, 107–8, 111
development 88–92
startup 91–92
SSC (Superconducting Super Collider) 126, 127
cancellation 129–30, 131
Standard Model 64–66
antimatter completing the 68
basic particles 64
extension of 155
free parameters in 81–82, 121
limitations 121
particles of 68
theories of 99
see also electroweak theory; QCD (Quantum Chromo Dynamics)
Steiger, Peter (son of Rudolf) 48
Steiger, Rudolf 47, 48, 49
strong interactions/forces 62, 65, 66
heavy force carriers 72
QCD theory 99
and QED 63
see also QCD (Quantum Chromo Dynamics); QED (quantum electrodynamics)
Stumpe, Bent, touchscreen proposal 151–52
superconductivity, BCS theory 76–77
SUSY (supersymmetry) theory 154–55
symmetry, concept of 73
see also broken symmetry (asymmetry); parity; spontaneous symmetry breaking
synchrotron radiation 98

tau particle 104, 117, 119
Taylor, Richard 71
Tevatron 3–4, 6, 92, 107–8
  run II 141
  top quark, discovery 108
  upgrade 121
Thatcher, Margaret 108, 144
Thomson, George (son of Joseph John) 40
Thomson, Joseph John 16, 25
  electrons, discovery 15
thought experiments
  Schrödinger's cat 24
  'spooky action at a distance' 24
'Three Coins in the Fountain' (Sinatra) 44
time reversal 74
Ting, Sam 115, 116
  J particle 102
  see also Richter, Burton
Tobias, Cornelius 146
Tomonaga, Sin-Itiro 80
top quarks
  discovery 104, 108
  mass 122
Touschek, Bruno 83, 84
Townsend, David 148
transition point (at PS) 57
  flipping 57–59
  Nescafé tin 58–59
triple symmetry, inviolability of 75
TRIUMF 27
  particle therapy center 146

UA1 and UA2 experiments, discovery of W and Z bosons 107–8
UNESCO (United Nations Educational, Scientific and Cultural Organization) 36
  Resolution No. 2.21 37
up quarks, colour charge (gluons) 64, 66, 117
upsilon meson 104

vacuum structure 78, 80
Valeur, Robert 45, 46, 47, 49
Van de Graaff, Robert Jemison 26
Veksler, Vladimir (synchrotron) 28
Veltman, Martinus 79, 80, 99
VEP-1 84, 92

W bosons (particles)
  carrier of weak force 67, 72, 79
  identification 107–8
  pairs 117, 121
  production at LEP II 113
weak interactions/forces 63, 65, 66
  mesons 62
  and QED 63
  understanding short range of 75–76
  see also mesons (mesotrons); QED (quantum electrodynamics); W bosons (particles); Z bosons (particles)
Weinberg, Steven 72, 75, 79
Weisskopf, Victor 61, 88
Widerøe, Rolf 26
Wilczek, Frank 100
Wilson, Robert 89, 107, 146
  defence of basic science 144–45

Yang, Chen Ning 74
Yukawa, Hideki 101
  strong force prediction 62, 66

Z bosons (particles)
  carriers of weak forces 67, 72, 79, 103, 105
  decay, detection 116–19, 121
  identification 107–8
  production at LEP I 113
Z line shape
  as measured by ALEPH experiment **118**
  models 118
  particle generations, finding number of 117–19
  role of neutrinos 118–19
  sensitivity to Z particle decay 118
Zweig, George, ace particles 69, 70